AF817979

Reviews of Environmental Contamination and Toxicology

VOLUME 252

More information about this series at http://www.springer.com/series/398

Reviews of Environmental Contamination and Toxicology

Editor
Pim de Voogt

Editorial Board
María Fernanda Cavieres, Valparaiso, Chile
James B. Knaak, Getzville, New York, USA
Annemarie P. van Wezel, Amsterdam, The Netherlands
Ronald S. Tjeerdema, Davis, California, USA
Marco Vighi, Madrid, Spain

Founding Editor
Francis A. Gunther

Volume 252

Springer

Coordinating Board of Editors

Dr. Pim De Voogt, *Editor*
Reviews of Environmental Contamination and Toxicology

University of Amsterdam
Amsterdam, The Netherlands
E-mail: w.p.devoogt@uva.nl

Dr. Erin R. Bennett, *Editor*
Bulletin of Environmental Contamination and Toxicology

Great Lakes Institute for Environmental Research
University of Windsor
Windsor, Ontario, Canada
E-mail: ebennett@uwindsor.ca

Dr. Peter S. Ross, *Editor*
Archives of Environmental Contamination and Toxicology

Vancouver Aquarium Marine Science Center
Vancouver, BC, Canada
E-mail: peter.ross@vanaqua.org

ISSN 0179-5953 ISSN 2197-6554 (electronic)
Reviews of Environmental Contamination and Toxicology
ISBN 978-3-030-30991-6 ISBN 978-3-030-30992-3 (eBook)
https://doi.org/10.1007/978-3-030-30992-3

© Springer Nature Switzerland AG 2020
This work is subject to copyright. All rights are reserved by the Publisher, whether the whole or part of the material is concerned, specifically the rights of translation, reprinting, reuse of illustrations, recitation, broadcasting, reproduction on microfilms or in any other physical way, and transmission or information storage and retrieval, electronic adaptation, computer software, or by similar or dissimilar methodology now known or hereafter developed.
The use of general descriptive names, registered names, trademarks, service marks, etc. in this publication does not imply, even in the absence of a specific statement, that such names are exempt from the relevant protective laws and regulations and therefore free for general use.
The publisher, the authors, and the editors are safe to assume that the advice and information in this book are believed to be true and accurate at the date of publication. Neither the publisher nor the authors or the editors give a warranty, express or implied, with respect to the material contained herein or for any errors or omissions that may have been made. The publisher remains neutral with regard to jurisdictional claims in published maps and institutional affiliations.

This Springer imprint is published by the registered company Springer Nature Switzerland AG
The registered company address is: Gewerbestrasse 11, 6330 Cham, Switzerland

Foreword

International concern in scientific, industrial, and governmental communities over traces of xenobiotics in foods and in both abiotic and biotic environments has justified the present triumvirate of specialized publications in this field: comprehensive reviews, rapidly published research papers and progress reports, and archival documentations These three international publications are integrated and scheduled to provide the coherency essential for nonduplicative and current progress in a field as dynamic and complex as environmental contamination and toxicology. This series is reserved exclusively for the diversified literature on "toxic" chemicals in our food, our feeds, our homes, recreational and working surroundings, our domestic animals, our wildlife, and ourselves. Tremendous efforts worldwide have been mobilized to evaluate the nature, presence, magnitude, fate, and toxicology of the chemicals loosed upon the Earth. Among the sequelae of this broad new emphasis is an undeniable need for an articulated set of authoritative publications, where one can find the latest important world literature produced by these emerging areas of science together with documentation of pertinent ancillary legislation.

Research directors and legislative or administrative advisers do not have the time to scan the escalating number of technical publications that may contain articles important to current responsibility. Rather, these individuals need the background provided by detailed reviews and the assurance that the latest information is made available to them, all with minimal literature searching. Similarly, the scientist assigned or attracted to a new problem is required to glean all literature pertinent to the task, to publish new developments or important new experimental details quickly, to inform others of findings that might alter their own efforts, and eventually to publish all his/her supporting data and conclusions for archival purposes.

In the fields of environmental contamination and toxicology, the sum of these concerns and responsibilities is decisively addressed by the uniform, encompassing, and timely publication format of the Springer triumvirate:

Reviews of Environmental Contamination and Toxicology [Vol. 1 through 97 (1962–1986) as Residue Reviews] for detailed review articles concerned with any aspects of chemical contaminants, including pesticides, in the total environment with toxicological considerations and consequences.

Bulletin of Environmental Contamination and Toxicology (Vol. 1 in 1966) for rapid publication of short reports of significant advances and discoveries in the fields of air, soil, water, and food contamination and pollution as well as methodology and other disciplines concerned with the introduction, presence, and effects of toxicants in the total environment.

Archives of Environmental Contamination and Toxicology (Vol. 1 in 1973) for important complete articles emphasizing and describing original experimental or theoretical research work pertaining to the scientific aspects of chemical contaminants in the environment.

The individual editors of these three publications comprise the joint Coordinating Board of Editors with referral within the board of manuscripts submitted to one publication but deemed by major emphasis or length more suitable for one of the others.

Coordinating Board of Editors

Preface

The role of *Reviews* is to publish detailed scientific review articles on all aspects of environmental contamination and associated (eco)toxicological consequences. Such articles facilitate the often complex task of accessing and interpreting cogent scientific data within the confines of one or more closely related research fields.

In the 50+ years since *Reviews of Environmental Contamination and Toxicology* (formerly *Residue Reviews)* was first published, the number, scope, and complexity of environmental pollution incidents have grown unabated. During this entire period, the emphasis has been on publishing articles that address the presence and toxicity of environmental contaminants. New research is published each year on a myriad of environmental pollution issues facing people worldwide. This fact, and the routine discovery and reporting of emerging contaminants and new environmental contamination cases, creates an increasingly important function for *Reviews.* The staggering volume of scientific literature demands remedy by which data can be synthesized and made available to readers in an abridged form. *Reviews* addresses this need and provides detailed reviews worldwide to key scientists and science or policy administrators, whether employed by government, universities, nongovernmental organizations, or the private sector.

There is a panoply of environmental issues and concerns on which many scientists have focused their research in past years. The scope of this list is quite broad, encompassing environmental events globally that affect marine and terrestrial ecosystems; biotic and abiotic environments; impacts on plants, humans, and wildlife; and pollutants, both chemical and radioactive; as well as the ravages of environmental disease in virtually all environmental media (soil, water, air). New or enhanced safety and environmental concerns have emerged in the last decade to be added to incidents covered by the media, studied by scientists, and addressed by governmental and private institutions. Among these are events so striking that they are creating a paradigm shift. Two in particular are at the center of ever increasing media as well as scientific attention: bioterrorism and global warming. Unfortunately, these very worrisome issues are now superimposed on the already extensive list of ongoing environmental challenges.

The ultimate role of publishing scientific environmental research is to enhance understanding of the environment in ways that allow the public to be better informed or, in other words, to enable the public to have access to sufficient information. Because the public gets most of its information on science and technology from internet, TV news, and reports, the role for scientists as interpreters and brokers of scientific information to the public will grow rather than diminish. Environmentalism is an important global political force, resulting in the emergence of multinational consortia to control pollution and the evolution of the environmental ethic. Will the new politics of the twenty-first century involve a consortium of technologists and environmentalists, or a progressive confrontation? These matters are of genuine concern to governmental agencies and legislative bodies around the world.

For those who make the decisions about how our planet is managed, there is an ongoing need for continual surveillance and intelligent controls to avoid endangering the environment, public health, and wildlife. Ensuring safety-in-use of the many chemicals involved in our highly industrialized culture is a dynamic challenge, because the old, established materials are continually being displaced by newly developed molecules more acceptable to federal and state regulatory agencies, public health officials, and environmentalists. New legislation that will deal in an appropriate manner with this challenge is currently in the making or has been implemented recently, such as the REACH legislation in Europe. These regulations demand scientifically sound and documented dossiers on new chemicals.

Reviews publishes synoptic articles designed to treat the presence, fate, and, if possible, the safety of xenobiotics in any segment of the environment. These reviews can be either general or specific, but properly lie in the domains of analytical chemistry and its methodology, biochemistry, human and animal medicine, legislation, pharmacology, physiology, (eco)toxicology, and regulation. Certain affairs in food technology concerned specifically with pesticide and other food-additive problems may also be appropriate.

Because manuscripts are published in the order in which they are received in final form, it may seem that some important aspects have been neglected at times. However, these apparent omissions are recognized, and pertinent manuscripts are likely in preparation or planned. The field is so very large and the interests in it are so varied that the editor and the editorial board earnestly solicit authors and suggestions of underrepresented topics to make this international book series yet more useful and worthwhile.

Justification for the preparation of any review for this book series is that it deals with some aspect of the many real problems arising from the presence of anthropogenic chemicals in our surroundings. Thus, manuscripts may encompass case studies from any country. Additionally, chemical contamination in any manner of air, water, soil, or plant or animal life is within these objectives and their scope.

Manuscripts are often contributed by invitation. However, nominations for new topics or topics in areas that are rapidly advancing are welcome. Preliminary communication with the Editor-in-Chief is recommended before volunteered review manuscripts are submitted. *Reviews* is registered in WebofScience™.

Inclusion in the Science Citation Index serves to encourage scientists in academia to contribute to the series. The impact factor in recent years has increased from 2.5 in 2009 to 7.0 in 2017. The Editor-in-Chief and the Editorial Board strive for a further increase of the journal impact factor by actively inviting authors to submit manuscripts.

Amsterdam, The Netherlands
August 2018

Pim de Voogt

Contents

Contributors

Bilal Ahmed Department of Microbiology, Faculty of Agricultural Sciences, Aligarh Muslim University, Aligarh, Uttar Pradesh, India

Aida Maria Benitez-Rexach Walden University, Minneapolis, MN, USA

Lieven Bervoets Laboratory of Systemic, Physiological and Ecotoxicological Research, Department of Biology, University of Antwerp, Antwerp, Belgium

Eliza Blicharska Department of Analytical Chemistry, Medical University of Lublin, Lublin, Poland

Abha Cherkani-Hassani Unit of Training and Research in Nutrition and Food Sciences, Faculty of Medicine and Pharmacy, University Mohammed V, Rabat, Morocco

Center of Doctoral Studies in Life Sciences and Health CEDoc-SVS, Faculty of Medicine and Pharmacy, University Mohammed V, Rabat, Morocco

Alexey Fedorenko Academy of Biology and Biotechnology, Southern Federal University, Rostov-on-Don, Russia

Jolanta Flieger Department of Analytical Chemistry, Medical University of Lublin, Lublin, Poland

Yanzheng Gao Institute of Organic Contaminant Control and Soil Remediation, College of Resources and Environmental Sciences, Nanjing Agricultural University, Nanjing, China

Imane Ghanname Team Research of Pharmacoeconomics and Pharmacoepidemiology, Laboratory of Pharmacology and Toxicology, Faculty of Medicine and Pharmacy, University Mohammed V, Rabat, Morocco

Faculty of Health Sciences, International University of Casablanca, Casablanca, Morocco

Andrey Gorovtsov Academy of Biology and Biotechnology, Southern Federal University, Rostov-on-Don, Russia

Fredrick Owino Gudda Institute of Organic Contaminant Control and Soil Remediation, College of Resources and Environmental Sciences, Nanjing Agricultural University, Nanjing, China

Bertrand Laratte Département de Conception, Industrialisation, Risque, Décision, Ecole Nationale Supérieure d'Arts et Métiers, Paris, France

Wanting Ling Institute of Organic Contaminant Control and Soil Remediation, College of Resources and Environmental Sciences, Nanjing Agricultural University, Nanjing, China

Juan Liu Institute of Organic Contaminant Control and Soil Remediation, College of Resources and Environmental Sciences, Nanjing Agricultural University, Nanjing, China

Saglara Mandzhieva Academy of Biology and Biotechnology, Southern Federal University, Rostov-on-Don, Russia

Patrick Meire Ecosystem Management Research Group (Ecobe), Department of Biology, University of Antwerp, Antwerp, Belgium

Tatiana Minkina Academy of Biology and Biotechnology, Southern Federal University, Rostov-on-Don, Russia

Nezha Mouane Unit of Training and Research in Nutrition and Food Sciences, Faculty of Medicine and Pharmacy, University Mohammed V, Rabat, Morocco

Department of Pediatric Hepatology Gastroenterology and Nutrition-P III, Rabat Children's Hospital, Rabat, Morocco

Javed Musarrat Department of Microbiology, Faculty of Agricultural Sciences, Aligarh Muslim University, Aligarh, Uttar Pradesh, India

Vishnu Rajput Academy of Biology and Biotechnology, Southern Federal University, Rostov-on-Don, Russia

Quaiser Saquib Zoology Department, College of Sciences, King Saud University, Riyadh, Saudi Arabia

Ritu Singh Department of Environmental Science, School of Earth Sciences, Central University of Rajasthan, Ajmer, Rajasthan, India

Mikhail Soldatov The Smart Materials Research Center, Southern Federal University, Rostov-on-Don, Russia

Svetlana Sushkova Academy of Biology and Biotechnology, Southern Federal University, Rostov-on-Don, Russia

Michael Gatheru Waigi Institute of Organic Contaminant Control and Soil Remediation, College of Resources and Environmental Sciences, Nanjing Agricultural University, Nanjing, China

Jian Wang Institute of Organic Contaminant Control and Soil Remediation, College of Resources and Environmental Sciences, Nanjing Agricultural University, Nanjing, China

Victor Wepener Water Research Group, Unit for Environmental Sciences and Management, North-West University, Potchefstroom, South Africa

Nico J. Wolmarans Water Research Group, Unit for Environmental Sciences and Management, North-West University, Potchefstroom, South Africa

Laboratory of Systemic, Physiological and Ecotoxicological Research, Department of Biology, University of Antwerp, Antwerp, Belgium

Bing Yang Institute of Organic Contaminant Control and Soil Remediation, College of Resources and Environmental Sciences, Nanjing Agricultural University, Nanjing, China

Endophytic Bacteria in in planta Organopollutant Detoxification in Crops

Michael Gatheru Waigi, Jian Wang, Bing Yang, Fredrick Owino Gudda, Wanting Ling, Juan Liu, and Yanzheng Gao

Contents

Michael Gatheru Waigi, Jian Wang, Bing Yang and Fredrick Owino Gudda contributed equally to this work.

M. G. Waigi · J. Wang · B. Yang · F. O. Gudda · W. Ling · J. Liu · Y. Gao (✉)
Institute of Organic Contaminant Control and Soil Remediation, College of Resources and Environmental Sciences, Nanjing Agricultural University, Nanjing, China
e-mail: 2017203065@njau.edu.cn; 2015203010@njau.edu.cn; 2016203038@njau.edu.cn; 2018203064@njau.edu.cn; lingwanting@njau.edu.cn; liujuan@njau.edu.cn; gaoyanzheng@njau.edu.cn

© Springer Nature Switzerland AG 2019

P. de Voogt (ed.), *Reviews of Environmental Contamination and Toxicology, Volume 252*, Reviews of Environmental Contamination and Toxicology Volume 252, https://doi.org/10.1007/398_2019_33

Abbreviations

2,4-D	2,4-Dichlorophenoxyacetic acid
ACCD	1-Aminocyclopropane-1-carboxylate deaminase
AMF	Arbuscular mycorrhizal fungi
CSIA	Compound-specific isotope analysis
DDBJ	DNA Data Bank of Japan
DDT	Dichlorodiphenyltrichloroethane
DEB_{OP}	Organopollutant-degrading endophytic bacteria
EMBL	European Molecular Biology Laboratory
EPS	Extracellular polymeric substances
GenBank®	A genetic sequence database maintained by the National Center for Biotechnology Information (NCBI) arm of the National Institutes of Health (NIH), USA
GFP	Green fluorescent protein
GST	Glutathione S-transferase
HRMS	High-resolution mass spectrometry
NGS	Next-generation sequencing
PAHs	Polycyclic aromatic hydrocarbons
PCBs	Polychlorinated biphenyls
PGP	Plant growth promotion/plant growth-promoting
POPs	Persistent organic pollutants
RefSeq®	Reference sequence, a database by the National Center for Biotechnology Information (NCBI) for single records of each natural biological resource (i.e., RNA, DNA or proteins) of living organisms, such as bacteria, viruses, and eukaryotes
SGS	Second-generation sequencing
TCE	Trichloroethylene
TGS	Third-generation sequencing
WHO	World Health Organization

1 Introduction

A wide spectrum of human-made organic compounds are utilized by man in various day-to-day applications. However, their anthropogenic release has given rise to organopollutants (also known as organic pollutants or pollutant organics), a contemporary environmental problem worldwide that results in crop contamination in farmlands, as man tries to meet the ever-increasing population demand and strive toward global agricultural sustainability (Sturz et al. 2000). These organopollutants include polyaromatic hydrocarbons (PAHs), polychlorinated biphenyls (PCBs), 2,4-dichlorophenoxyacetic acid (2,4-D), dichlorodiphenyltrichloroethane (DDT), dibenzofurans, halogenated aromatic compounds, nitroaromatic compounds, phenolic compounds, and various pesticides, which are hazardous to the environment. Organopollutants and their degraded forms find their way in farmland soils, irrigated groundwater, or ambient air, which exacerbates the crop contamination risk (Feng et al. 2017a). In vegetables and fodder crops, their transfer to various plant parts via phytouptake from soil matrices or their intake from the surrounding atmosphere has been well studied (Binet et al. 2000; Bittsánszky et al. 2011; Chaîneau et al. 1997; Chiou et al. 2001; Collins et al. 2006; Collins 2008; Gao and Zhu 2004; Hellström 2004; Kacálková and Tlustoš 2011; McKone and Maddalena 2007; Zhang et al. 2017). This leads to their transfer within trophic levels, which pose adverse environmental and human health risks (Ashraf 2017). A schematic overview of POP entry in crops is seen in Fig. 1.

The presence and entry of organopollutant xenobiotics in crops negatively influence various living entities. They are toxic, persistent in the environment, bioaccumulating, and biomagnifying (Wang et al. 2017a), hence may have far-reaching consequences for living organisms dependent on plants (Pereira

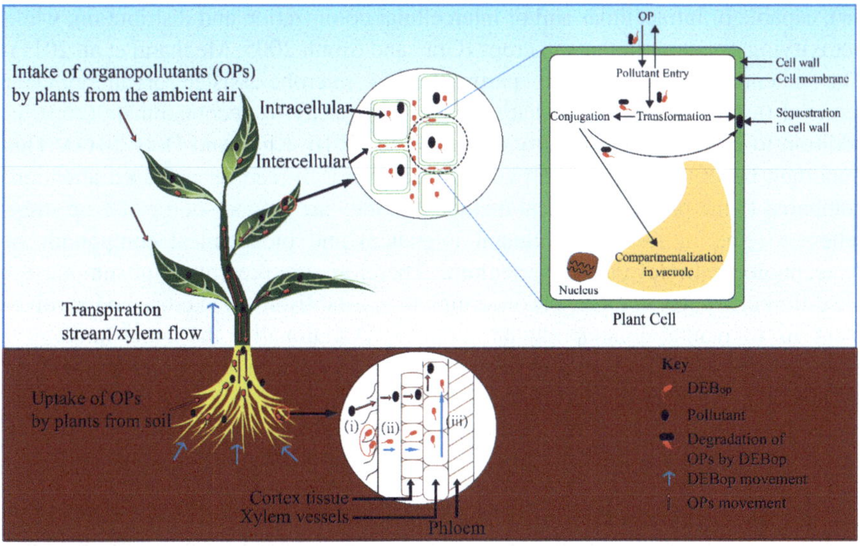

Fig. 1 DEB_{OP} colonization, distribution, and mode of action in POPs reduction in crops: the interplay between endophytic degradation and the green liver concept. Crops would selectively allow members of certain soil bacterial classes to intercellularly or intracellularly colonize their internal tissues. Through active or passive root penetration strategies in colonization, DEB_{OP} may enter at the root epidermal tissues before reaching the root cortical zone. Following their entry at the root cortex, colonization at the lumen of xylem vessels by the DEB_{OP} occurs before their traverse movement to the aerial parts of the plant via xylem flow. After the phytouptake of organopollutant xenobiotics, crops induce the green liver concept of plant metabolism (i.e., transformation, conjugation, and compartmentalization) in the root and shoot endosphere. The selectively colonized DEB_{OP} (intercellular and intracellular) may induce microbial degradative activity, before or during phytotransformation. This interplay enhances the mechanisms of plant-microbe interactions utilized by DEB_{OP}

2014). According to the World Health Organization (WHO), the impact and fate of organopollutants, even in low concentrations, may lead to the alteration of the immune system, neurobehavioral impairment, reproductive disorders, genotoxicity, increased congenital disabilities, endocrine disruption, and increased cancer risk (Nadal and Domingo 2013; Li et al. 2006; WHO 2018). This shows that the presence of organic contaminants in various crops above their toxic limits poses a global concern, as the contamination risk leads to bioaccumulation in humans at the top of the food chains (Nadal and Domingo 2013).

Crops interact with numerous microorganisms (bacteria and fungi) as part of their evolved adaptation strategy in organopollutant-stressed farmlands (Wang et al. 2010). Among these plant-beneficial microbes support various mechanisms such as microbe-assisted phytoremediation, plant growth promotion, synthesis of essential antifungal/antibiotic substances for phytopathogens, tolerance to various environmental stresses, and microbe-assisted organopollutant removal (Hallmann et al. 1997; Wu et al. 2009). The latter approach has been utilized by beneficial microorganisms called *organopollutant-degrading endophytic bacteria* (DEB_{OP}), a range of nonpathogenic bacteria from various taxa (originating from the soil or surrounding air), capable of intracellular and/or intercellular colonization and distribution, while detoxifying organopollutants in crops (Gray and Smith 2005; Megharaj et al. 2011). The salient advantage of this partnership in microbe-assisted organopollutant removal (a type of phytoremediation) is their capacity to decontaminate crops, in addition to inducing crop fertility (Eevers et al. 2016; Khan and Doty 2011). This important benefit has made the two-pronged approach receive increased attention, compared to the physicochemical methods, as they are a good source of crop stress relievers (e.g., in salt and pollutant tolerance) and biochemical compounds of biotechnological relevance in agriculture. Therefore, the present script summarizes the different aspects of DEB_{OP}, concerning their establishment in crops and involvement in in planta organopollutant removal. Additionally, the biotechnological advances, benefits, and challenges in DEB_{OP} application in agriculture will be reassessed. This microbial technology will provide a further understanding of the prospects and outcomes of the coexistence and joint roles of DEB_{OP} and their host partners in crop protection from organopollutants while enhancing plant productivity and fertility.

2 The Concept and Features of DEB_{OP} in the Elimination of Organopollutants

In the context of crops and their associated bacteria, there are three types of symbiotic relationships (Newton et al. 2010), which literally means the "living together" of an extensive cortège of microbial organisms, irrespective of a beneficial, neutral, or detrimental outcome (Brader et al. 2017). These symbiotic interactions may either be mutualistic (Perotti 1926) or pathogenic/parasitism (Gagne et al. 1989;

Hollis 1949, 1951; Smith 1911; Tervet and Hollis 1948), and commensalistic (De Bary 1879; Hentschel et al. 2000). Bacterial endophytes, in general, are either phyllospheric or rhizospheric in origin (Arslan et al. 2017; Kloepper et al. 1992). The focus of this review is on mutualistic endophytic bacteria emanating from the soil (i.e., rhizospheric and free-living bacteria), the site of intense interactions between plant and their bacterial partners, and a determinant of plant microbiota (Chanway 1996). First discovered by Victor Galippe (1887a, b), endophytic bacteria-plant interactions have been extensively studied for their diverse roles, especially in plant growth promotion, biocontrol, pathogenicity, remediation, and stress tolerance (Compant et al. 2005, 2010, 2012; Fester et al. 2014; Hardoim and van Elsas 2013; Hardoim et al. 2008; McCully 2001; Reinhold-Hurek and Hurek 2011; Zinniel et al. 2002). These nonpathogenic microorganisms have been found in virtually every plant (Bacon and White 2000; Hallman 2001; Phillips et al. 2008). However, only about 1% have been identified for further analysis (Eevers et al. 2015d). The low number of investigated bacterial endophytes indicates a research gap that highlights the impact of endosymbiotic bacteria-plant associations in crops, particularly in organopollutant reduction.

Although the foundation of the endophyte research can be dated back to the nineteenth century, investigations on this field of research have only recently thrived, churning an expanding body of literature (Compant et al. 2016, 2019). A few recent developments in this topic have arisen, especially in terms of the bacterial taxa involved and their interactions with crops in contaminant removal (Fester et al. 2014; Ryan et al. 2008; Weyens et al. 2009a). This has led to the discovery of members of bacterial taxonomic groups such as *Pseudomonas* (Germaine et al. 2009), *Enterobacter* sp. (Eevers et al. 2015a; Sheng et al. 2008; Yousaf et al. 2011), *Sphingomonas* sp. (Feng et al. 2017a), *Bacillus* sp. (Eskandary et al. 2017), *Burkholderia* sp. (Compant et al. 2008; Mitter et al. 2013a, b), *Serratia* sp. (Dong et al. 2014; Sessitsch et al. 2011), and *Paenibacillus* (Zhu et al. 2016a). These bacterial endosymbionts have been isolated and utilized in organopollutant decontamination as carbon and energy sources (see Table 1). The discovery of endophytic bacteria in crops gives a clear indication of their diversity in microbial communities in crop growth and adaptability to organically polluted ecosystems.

Soil microorganisms in nature can penetrate healthy plant tissues after their establishment in the rhizosphere (Compant et al. 2010; Galippe 1887a, b; McCully 2001). As DEB_{OP} belong to these soil microbes, they have the following four attributes which distinguish them from a plethora of microbes from either plants or soils (Compant et al. 2010; Kandel et al. 2017; Lumactud et al. 2016). Initially, DEB_{OP} are capable of entering the root endosphere of their hosts, once the rhizoplane is colonized (Compant et al. 2010; Lumactud et al. 2016). Then, the endosymbiotic bacteria induce a distribution strategy (microbial proliferation and movement in crops) within the various plant bodies (Brader et al. 2014; Feng et al. 2017a; Santoyo et al. 2016). Afterward, the endophytic microorganisms stimulate plant growth-promoting effects of survival in contaminated environments (Compant et al. 2010; Glick 2012, 2003, 2015; Lumactud et al. 2016; Sheng et al. 2008). Finally, the bacterial endosymbionts portray inherent traits (e.g., abundance and

Table 1 Classical examples of endophytic bacteria and crops in microbe-assisted organopollutant removal

POPs	Endophytic bacterial species	Plant species	Main attributes of plants	Main attributes of DEB_{OP}	References
Chlorpyrifos	*Sphingomonas* sp. strain HJY	*Allium tuberosum* Rottl. ex Spreng	No direct attributes in plants	Higher degradation of CP inside the inoculated plants than in the uninoculated plants	Feng et al. (2017a)
16 USEPA priority PAHs	Composite of 8 endophytic bacteria in PAH reduction (EB_{PAH})	*Brassica chinensis* L. and *Brassica campestris* L.	No direct attributes in plants. Other attributes include reducing the health risk from dietary exposure (of vegetables grown in PAH-contaminated areas)	Health risk reduction posed by vegetables contaminated with PAHs, due to high PAH removal efficiency in edible vegetable parts	Wang et al. (2017b)
Pyrene	*Serratia* sp. PW7	*Triticum aestivum* L. cv. Yangmai-16	A significant increase in growth of seedlings from root and shoot groups (TR and TS, respectively). Shoots grew by 18% after 4 days and roots by 29–31% in 4–8 days	Reduction of pyrene content in vitro and in vivo	Zhu et al. (2017a, b)
Naphthalene	*Pseudomonas putida* VM1441(pNAH7)	*Pisum sativum* var. Early onward	High seed germination and plant transpiration rates observed when exposed to naphthalene	Efficient plant colonizer, with biofilm and microcolony formation in both the rhizosphere and internal root tissues. High (40%) naphthalene degradation rates	Germaine et al. (2009)
2,4-Dichlorophenoxyacetic acid (2,4-D)	*Pseudomonas putida* 1450	*Pisum sativum*	Increase in, and maintenance of, growth and of biomass, as well as an increase in plant uptake of 2,4-D from the soil	Microbe-assisted phytodegradation of 2,4-D	Germaine et al. (2006)

Phenanthrene	*Pseudomonas* sp. Ph6-gfp	*Lolium multiflorum* Lam.	No direct attributes in plants	Ph6-gfp decreased concentration, accumulation, and translocation factors of phenanthrene	Sun et al. (2014a), (2015a, b), (2018)
Chlorpyrifos	*Pseudomonas aeruginosa* strain RRA, *Bacillus megaterium* strain RRB, *Sphingobacterium siyangensis* strain RSA, *Stenotrophomonas pavanii* strain RSB, and *Curtobacterium plantarum* strain RSC	*Oryza sativa* L.	Observed plant growth promotional traits, including indole acetic acid and siderophore production, secretion of phosphate solubilization and 1-aminocyclopropane-1-carboxylate deaminase	Phytodegradation of CP led to the reduction of the residue of up to 80% in the infested rice grains, compared to the controls	Feng et al. (2017b)
2,2-Bis (p-chlorophenyl)-1,1-dichloroethylene (DDE)	*Sphingomonas taxi* UH1, *Methylobacterium radiotolerans* UH1, and *Enterobacter aerogenes* UH1	*Cucurbita pepo*	Presence of genes for biodegradative enzymes implicates the association with DDE, PAHs, and nitriloacetate degradation	The presence of various PGP genes, indicating growth and survival in POP-contaminated environments. In field trials, the inoculated consortium of the 3 strains induced enhanced phytoaccumulation and degradation in zucchini plants	Eevers et al. (2015a, b, c), (2018)
Phenanthrene	*Massilia* sp. Pn2	*Triticum aestivum* L. cv. Yangmai-16	Phenanthrene contamination of wheat was reduced	Pn2 changed the bacterial community structure and enhanced the cell counts, diversity, and richness of some endophytic bacteria in wheat	Liu et al. (2017)
Naphthalene, fluorene, phenanthrene, fluorene, pyrene, and benzo[*a*] pyrene	*Paenibacillus* sp. PHE-3	*Plantago asiatica* L. isolate was inoculated into *Triticum aestivum*	No direct attributes in plants	99.9% degradation after 84 h cultivation with additional glutamate. Presence of catechol	Zhu et al. (2016a)

(continued)

Table 1 (continued)

POPs	Endophytic bacterial species	Plant species	Main attributes of plants	Main attributes of DEB_{OP}	References
		L. cv. Ningmai-13 and *Zea mays* L. cv. Xiyu-3		2,3-dioxygenase activity during degradation. Effective degradation of 2- and 3-ringed PAHs by strain PHE-3 through co-metabolism	
Diesel	*Pseudomonas* sp. strain ITRI53 and *Rhodococcus* sp. strain ITRH43	Isolated from *Lolium multiflorum* Lam.	No direct attributes in plants	DEB_{OP} showed higher expression of alkane monooxygenase (*alkB*) genes	Andria et al. (2009), Yousaf et al. (2010a, b)
Diesel	*Enterobacter ludwigii*	*Lolium multiflorum* var. Taurus, *Lotus corniculatus* var. Leo, and *Medicago sativa* var. Harpe	Biomass production was significantly higher in inoculated treatments than in uninoculated contaminated treatments (especially for the second harvest than in the first harvest)	DEB_{OP} portrayed efficient colonization in the rhizosphere and plant interior; also, the expression of the CYP153 gene in all plant parts was observed, indicating an active role of DEB_{OP} in diesel degradation, in association with plants	Yousaf et al. (2011)
Pyrene (PAH)	*Enterobacter* sp. 12J1	Isolated from *Alternanthera philoxeroides*, *Allium macrostemon* Bunge, *Dendranthema indicum*, and *Echinochloa crus-galli*	Plant growth promotion of wheat and maize, which induced plant-promoted biodegradation of pyrene	Colonization of the interior plant tissues (root and stem) of wheat and maize, leading to the increased rate of pyrene removal by 43–65% in the live	Sheng et al. (2008)

				bacterium-inoculated planted soils	
Phenanthrene and pyrene (PAHs)	*Bacillus* sp., *Pantoea* sp., *Pseudomonas* sp., *Arthrobacter* sp., *Pedobacter* sp., and *Delftia* sp.	*Lolium multiflorum* Lam.	No direct attributes in plants	Increased due to exposure to phenanthrene and pyrene (PYR), while the abundance was increased by individual PAH exposure	Zhu et al. (2016b)
Toluene	*Achromobacter xylosoxidans* F3B	*Chrysopogon zizanioides* (vetiver grass)	Inoculation of *A. xylosoxidans* F3B to *Arabidopsis thaliana* resulted in a significant increase in root length, fresh weight in catechol and phenol-added agar	F3B was able to utilize 30 different substrates. F3B can also help the plant tolerate lethal concentrations of aromatic compounds and decrease catechol- and phenol-induced phytotoxicity, as well as reducing evapotranspiration to the leaves	Ho et al. (2009), (2012), (2013)
Hexachlorocyclohexane (HCH)	*Rhodococcus erythropolis* ET54b and *Sphingomonas* sp. D4	*Cytisus striatus* (Hill) Rothm. (Portuguese broom)	Substrates planted with *C. striatus* showed a higher dissipation of HCH isomers, due to inoculation of this plant species with two bacterial strains in combination (*Rhodococcus erythropolis* ET54b and *Sphingomonas* sp. D4) led to an increase in the shoot and root biomass	Inoculation of this plant species with two bacterial strains in combination protected the plants against the toxic effects of the contaminant in the root and shoot endosphere. The *Rhodococcus* had PGP traits such as phosphate solubilizer, as well as a biosurfactant and IAA producer. The sphingomonad is a HCH-degrader	Becerra-Castro et al. (2011), (2013a, b)

(continued)

Table 1 (continued)

POPs	Endophytic bacterial species	Plant species	Main attributes of plants	Main attributes of DEB_{OP}	References
Diesel	*Burkholderia phytofirmans* PsJN	*Lolium multiflorum* Lam.	Application of the plant growth-promoting and hydrocarbon-degrading bacterium *B. phytofirmans* PsJN with different inoculation methods improved shoot length (8–31%) and root (2–26%) and shoot (4.4–38%) weight as compared to non-inoculated plants. Maximum plant biomass (16.6 g pot^{-1}) was obtained with soil inoculation	Root (rhizosphere and endosphere) and shoot (interior) colonizer. Hydrocarbon-degrading endophyte Based on its large genome, it can successfully colonize such a wide variety of plant species	Afzal et al. (2013b), Mitter et al. (2013a)
Petroleum hydrocarbons	29 strains from 11 different genera, with most of them belonging to *Pseudomonas*, *Bacillus*, and *Curtobacterium* genera	*Lolium perenne* L.	No direct attributes in plants	The endophytic bacteria confirmed various PGP properties, e.g., IAA, ACCD, NH4, siderophore, HCN production, as well as P solubilization	Kukla et al. (2014)
Phenanthrene	*Massilia* sp. Pn2	*Lolium multiflorum* Lam.	Plant growth promotion by increasing the fresh weight and dry weight, as well as ryegrass height and root length	Phenanthrene concentrations in ryegrass roots and shoots were reduced by 54% and 57%	Yousaf et al. (2011)
n-Hexadecane, diesel fuel, and PAH	*Pseudomonas* spp., *Brevundimonas*, *Pseudomonas rhodesiae*	*Lolium perenne* L., *Medicago sativa* L. var. Rambler, and *Festuca rubra* L., *Agropyron elongatum* P. Beauv,	Maintenance of distinct, heterogeneously distributed endophytic microbial populations	Heterogeneous endophytic microbes in plants may impact the ability of plants to promote the degradation of specific hydrocarbons	Phillips et al. (2008)

		Elymus angustus Trin., *Puccinellia nuttalliana* (Schult.) A.S. Hitchc			
Pyrene	*Staphylococcus* sp. BJ06	*Lolium multiflorum* Lam.	Plant growth promotion was observed, with increases in fresh weight, dry weight, as well as rye-grass height and root length	Pyrene concentrations in roots and shoots were reduced by 31.01% and 44.22%, respectively	Sun et al. (2014b)
Toluene	*Burkholderia cepacia* strain FX2	*Zea mays* and *Triticum durum*	Plant growth promotion, as observed from increased fresh and dry weights	*Burkholderia cepacia* FX2 which was found in inoculated plants, induced the reduced evapotranspiration of toluene into the air	Wang et al. (2010)
Toluene	*Burkholderia cepacia* G4 plasmid conjugated into *B. cepacia* strain BU0072, a derivative of *B. cepacia* L.S.2.4	Yellow lupine (*Lupinus luteus* L.)	A phytoprotective effect was seen when yellow lupine was inoculated with *B. cepacia* BU0072, as it reduced toluene phytotoxicity in the group treated with 100 mg/l or lower. However, plants exposed to toluene concentrations higher than 100 mg/l died within 1 week. A clear phytoprotective effect of the transconjugant *B. cepacia* VM1330 was observed on the growth of yellow lupine at all tested toluene concentrations	*Burkholderia cepacia* strain BU0072 is a genetically engineered endophytic bacteria which improved the efficiency of in planta degradation of volatile water-soluble organic xenobiotics such as toluene	Barac et al. (2004)

(continued)

Table 1 (continued)

POPs	Endophytic bacterial species	Plant species	Main attributes of plants	Main attributes of DEB_{OP}	References
Nickel-TCE co-contamination	*Burkholderia cepacia* VM1468 (conjugated with pTOM-Bu61 plasmid and *ncc-nre* Ni resistance/sequestration system)	Yellow lupine (*Lupinus luteus* L.)	A decrease in Ni and TCE phytotoxicity, as seen by the 30% increased root biomass, while up to 50% decreased antioxidative defense enzymatic activities in the roots was observed. In contrast, the inoculation with gfp-labeled *P. putida* W619 had a negative effect on plant health and growth	A 90% reduction in TCE evapotranspiration was observed, with a 5 times higher Ni phytouptake	Weyens et al. (2010a)
Nickel-TCE co-contamination	*Pseudomonas putida* strain W619	Hybrid poplar trees [Populus deltoides × (Populus trichocarpa × Populus deltoides) cv. Grimminge]	*P. putida* W619 (wild-type) activity resulted in remarkable growth promotion, decreased activities of antioxidative defense-related enzymes, and reduced stomatal resistance, all indicative of improved plant health and growth	The isolate induced in planta TCE degradation and nickel phytotoxicity, which aided plant	Weyens et al. (2009a, b, c), (2010b), (2012)
Chlorpyrifos (CP)	*yfp*-tagged *Mezorhizobium* sp. HN3	Ryegrass (*Lolium multiflorum* var. Taurus)	Colonization of the endophyte at the root endospheric and rhizospheric level, which enhanced plant growth	Degradation of CP and its metabolite TCP (3,5,6-trichloro-2-pyridinol), leading to a marked reduction of CP residues in plants roots and shoots	Jabeen et al. (2016)

				vegetated in inoculated soil, an attribute of the efficient root colonization of HN3*yfp*	
Diesel (1.25%, w/w)	*Rhizobium pisi* ER50 and *Bradyrhizobium japonicum* ER33 are endophytic bacteria; *Streptomyces costaricanus* RP92 *Pseudomonas* sp. 12 *Pantoea ananatis* 105 *Bacillus licheniformis* 255	*Cytisus striatus* L. and *Lupinus luteus* L.	A significant plant growth was observed in *L. luteus* in the consortium of strains than in individual strains, while individual inoculation treatments were more effective for *C. striatus*. In addition, a marked improvement of *L. luteus* in diesel-contaminated soil was seen, in the presence of PGP strains, leading to a twofold or higher increase in plant biomass. Only a subset of siderophore-producing and P-solubilizing bacterial strains that resulted in significantly modification of Fe or P concentrations in leaf tissues	Inoculants did not induce significant changes in plant nutritional status, with the exception of a few siderophore-producing and P-solubilizing bacterial strains, resulting in significant modification of Fe/P concentrations in leaf tissues. Inoculants did not cause significant changes in enzyme activities in perlite experiments; however, they significantly reduced oxidative stress in contaminated soils suggesting an improvement in plant tolerance to diesel	Balseiro-Romero et al. (2017)

Classic examples of DEB_{OP} applications, their isolation from/inoculated into plants, and the beneficial attributes emanating from both the hosts (plants) and the endosymbionts (DEB_{OP}). *ACCD* 1-aminocyclopropane-1-carboxylate deaminase, *NH_4* ammonia production, *HCN* hydrogen cyanide, *IAA* indole acetic acid, *P solubilization* phosphate solubilization, *PGP* plant growth promotion/promoting, *PAHs* polycyclic aromatic hydrocarbons, and *USEPA* United States Environmental Protection Agency expression of catabolic pollutant-degrading genes, extracellular endophytic-derived enzymes, extracellular polymeric substances, and biosurfactants) that assist in the mineralization or partial degradation of organic pollutants within the host crops (Feng et al. 2017a, b, c, d; Germaine et al. 2009). The four distinct features mentioned are essential in DEB_{OP} establishment, survival, and organopollutant reduction in crops, which will be critically examined in the subsequent sections of this review.

3 Colonization and Distribution Strategies of DEB_{OP}

Prior to the endophytic colonization (i.e., the entry, growth, and multiplication of endophyte populations within the host plant) of soil-borne bacteria to form DEB_{OP} in the endorhizosphere, rhizobacteria occur and proliferate the ectorhizosphere and rhizoplane regions via the rhizospheric effect (Compant et al. 2010; Fester et al. 2014; Liu et al. 2018; Reinhold-Hurek and Hurek 2011; Reinhold-Hurek et al. 2015). As detailed in Fig. 1, the microcolony formation of DEB_{OP} in crops involves an array of specific traits of *endophytic competence* (the ability of a bacterium to successfully establish in various plant bodies) (Compant et al. 2010). Some essential features such as motility, entry (via plant cell-wall degradation ability), and reactive oxygen species (ROS) scavenging are significant in their adaptability in crops (Liu et al. 2017). Also, the plant-driven selection of soil microorganisms occurs, as rhizodeposition from plants greatly influences the surrounding soil and its microflora (Hartmann et al. 2009), as well as the entry of endophytic bacteria in plants. After their entry in epidermal tissues, endophytic bacteria may either be localized in plant tissues such as the root cortex or xylem vessels or spread systematically throughout the plant via the vascular system or the apoplast (Thijs et al. 2018). As the plant microbiome research into endophytic colonization has gained considerable attention, the mechanisms employed by plants in recruiting endophytes are mostly unknown (Kandel et al. 2017). The studies into the interactive and versatile nature of a DEB_{OP} toward its host may give an in-depth understanding into the survival of a crop organopollutant-contaminated farmlands.

A compelling interest in bacterial endophytic research has recently bloomed (Hardoim et al. 2015; Lodewyckx et al. 2002; Rylott 2014; Weyens et al. 2009c; Zhu et al. 2014), particularly in the establishment and survival of the endosymbionts in their hosts (Compant et al. 2010; Gaiero et al. 2013). The abovementioned developments that have unearthed the endophytic traits, lifestyle, and activity, plus their associated impact on crops, are presented in Table 1. Some of these investigations employed the autofluorescence protein (AFP) technique, which use fluorescent proteins like green fluorescent protein (GFP), enhanced cyan (ECFP), enhanced green (EGFP), enhanced yellow (EYFP), red fluorescent protein (DsRed), and *mCherry* (Lagendijk et al. 2010). AFP technique is a relevant molecular identification method of illuminating biofilm formations at the rhizosphere level, as well as microcolony formations formed in the endosphere level in plants (Germaine et al.

2009; Lagendijk et al. 2010; Bloemberg et al. 2000). The observation of plant-microbe interactions using advanced microscopy methods such as electron microscopy, confocal laser scanning microscopy, Raman spectroscopy, super-resolution microscopy, and atomic force microscopy would help in understanding the endophytic bacteria-plant interactions as whole cells and at micro-scale levels (Cardinale and Berg 2015). The combination of such high-tech microscopic tools with autofluorescence tagging would provide new insights into the modes and sites of entry of endophytic bacteria, as well as evidence of an endophytic lifestyle and microbe-plant interactions (Cardinale 2014; Ryan et al. 2008).

The use of GFP technique to visualize endophytic bacteria under fluorescence microscopy revealed the biofilm and microcolony formations of the flagellated *Pseudomonas* sp. Ph6-*gfp* in the roots, stems, and leaves in PAH-contaminated Italian ryegrass (*Lolium multiflorum* Lam.) plants, compared to Ph6-*gfp*-free plants, an indication of the occurrence of active root penetration and distribution in the Ph6-*gfp* strains in ryegrass plants (Sun et al. 2014a, b, 2015a). The endophytic bacterium *Pseudomonas putida* VM1441 (pNAH7) colonized at the root level (rhizoplane, epidermal intercellular spaces, the cortex, and near the vascular tissue cells) of the pea plants (*Pisum sativum* L.) during naphthalene removal (Germaine et al. 2009). However, while inoculating the isolate *Pseudomonas putida* VM1450 in the same plant type during 2,4-D detoxification, rhizospheric and endospheric colonization (in root and stem tissues) were observed in their biofilm and microcolony visualizations using an epifluorescence microscope (Germaine et al. 2006, 2009). This ascending movement by DEB_{OP} using the xylem flow corroborated a study by Zhu et al. (2017a), where they noted that the isolate *Serratia* sp. PW7 utilized the transpiration stream in wheat (*Triticum aestivum* L.) crops, in their various tested inoculation methods. Furthermore, the subcellular distribution of the endophytic colonizer *Pseudomonas* sp. Ph6-*gfp* between the cell walls, intercellular spaces, vascular bundle tracheas, and cell organelles of pak choi (*Brassica chinensis* L.) seedlings were visualized using fluorescence microscopy (Sun et al. 2018). This subcellular colonization and distribution altered the subsequent phenanthrene biotransformation in pak choi leaves, leading to a sharp increase in cell numbers observed with increasing inoculation time (Sun et al. 2018). This is in line with some reviews, which indicate that endophytic bacteria primarily colonize inter- and/or subcellularly in plants (Hardoim et al. 2008; Thijs et al. 2018). Going by the findings by Germaine and his colleagues (Germaine et al. 2006, 2009), pseudomonad isolates will move further up a plant under PAH stress, than under 2,4-D, which indicates that the more soluble organopollutants such as two- and three-ringed PAHs allow for further distribution up a plant (rhizospheric and endospheric colonization throughout the plant), as opposed to 2,4-D (rhizospheric and root endospheric colonization, ending near the vascular tissues). The findings in the mentioned investigations shed light on the influence of certain organopollutants on the traits of endophytic competence and the inherent crop behavior under study.

Bacterial endophytic performance in a crop varies with the inoculation type, which has an impact on the in planta detoxification performance. Afzal et al. (2012) drew attention to this feature, whereby the inoculation method plays a vital role in the biomass growth and hydrocarbon degradation in Italian ryegrass (*Lolium multiflorum* var. Taurus). They found out that the performance strength of the strains (*Pseudomonas* sp. strain MixRI75 and *Pantoea* sp. strains ITSI10 and BTRH79) correlated with the colonization efficiency in the plants (soil inoculation>seed soaking). This finding corresponded to the results obtained by Zhu et al., where they noted that the inoculation type (seed soaking, seedling root soaking, and leaf spraying) varied according to the types of bacterial endophyte and the host crop (Zhu et al. 2017a). This shows that biotic properties may dictate the entry of DEB_{OP} via seeds, roots, or leaf, along with the type and uptake of the organopollutant present in the plant (Sun et al. 2015b; Zhu et al. 2017a). Overall, the use of autofluorescence techniques, microscopy, and other microbial technologies will provide valuable insights into the spatial dynamics of host-microbe interactions.

4 Plant Growth Promotion by DEB_{OP} in Organopollutant Degradation in Crops

Apart from their colonization and distribution in crops, organopollutant-degrading bacterial endophytes have also been detected for their role in plant growth promotion (PGP) (Etesami et al. 2015; Rashid et al. 2012). DEB_{OP} may foster plant growth by mimicking similar mechanisms as their plant partners in their survival and in planta organopollutant degradation activities in crops grown in organic-contaminated environments (Ali et al. 2017; Chanway 1996; Gaiero et al. 2013; Gamalero and Glick 2011; Olanrewaju et al. 2017). The direct and indirect plant growth-promoting mechanisms assist crops in growth, development, and phytoprotection in polluted ecosystems due to the PGP traits working in tandem with organopollutant detoxification (Chaturvedi et al. 2016). Plants also stimulate enhanced microbial activity in the endosphere (Begum and Tamilselvi 2016), which mirrors the rhizospheric effect in rhizospheric microbes (Ahemad and Kibret 2014; Bhattacharyya and Jha 2012; Gupta et al. 2015; Kloepper and Ryu 2006; Vejan et al. 2016). The elevated activity induced through the supply of oxygen, vitamins, and organic compounds (such as plant polyphenols) is vital for the selective bacterial enrichment and specific induction of catabolic genes (Afzal et al. 2013b; Harish et al. 2009; Pirttilä et al. 2004; Siciliano and Germida 1998). The PGP mechanisms by DEB_{OP} are crucial to both partners in their symbiotic relationship, thereby enhancing their host's survival in contaminated farms.

Few studies have pointed toward the application of bacterial endophytes (such as DEB_{OP}) that have both plant growth-promoting and pollutant-degrading attributes, therefore becoming superior strains compared to those possessing only one of these traits (Glick 2010, 2014; Glick and Stearns 2011). From the well-known PGP stimulating *Enterobacteriaceae* family, the bacterial strains *Enterobacter ludwigii* ISI10-3 and BRI10-9 showed colonization in root and shoot endosphere with 1-aminocyclopropane-1-carboxylate deaminase (ACCD) activity, thus revealing a tendency toward a more efficient plant beneficial and hydrocarbon degradation (Yousaf et al. 2010b, 2011). In another investigation, using a genome survey of the plant growth-promoting bacterium *Burkholderia phytofirmans* PsJN, the strain possessed a wide array of genes encoding for organopollutant degradation, as well as various plant growth-promoting and drought-resistant characteristics (Mitter et al. 2013b). The endophytic *Pantoea* spp. isolated from the microbial community within herbaceous plants from petroleum hydrocarbon-contaminated and non-contaminated soils displayed PGP characteristics, such as phosphate solubilization, indole-3-acetic acid production, and presence of ACCD (Lumactud and Fulthorpe 2018). Some pollutant-degrading epiphytic bacteria have also been found to be nitrogen fixers (Afzal et al. 2014; Ali et al. 2012). However, this trait has not been validated in research investigations involving DEB_{OP}. These investigations show that genes encoding for PGP traits are a distinguishable feature of DEB_{OP} establishment in organopollutant-contaminated crops.

Bacterial endophytes such as DEB_{OP} may act alone or in a composite with other organopollutant degraders in the plant microbiome. In a study by Wang et al. (2017b), a composite of eight PAH-degrading endophytic bacteria originating from *Sphingobium*, *Mycobacterium*, *Massilia*, *Paenibacillus*, and *Pseudomonas* taxa was used, which demonstrated the consortial application of DEB_{OP} in improving plant growth and degrading the PAHs in the Chinese cabbage (*Brassica chinensis* L.) and pak choi (*Brassica campestris* L.). The inoculation of the well-researched *Achromobacter xylosoxidans* F3B to *Arabidopsis thaliana* resulted in a significant increase in root length, as well as reducing catechol- and phenol-induced phytotoxicity while allowing lethal concentrations of aromatic compounds (Ho et al. 2009, 2012, 2013). The exploitation of bacteria from different taxonomical groups is beneficial to the plant, as seen in the combined inoculation of the PGP strain *Rhodococcus erythropolis* ET54b and the HCH-degrading sphingomonad *Sphingomonas* sp. D4 in *Cytisus striatus* plants (Becerra-Castro et al. 2011, 2013a, b). As shown in Table 1, several endophytic bacteria have been investigated for exhibiting diverse plant growth-promoting activities, which increase a crop's adaptability and versatility when encountering in planta contamination. These include siderophore production, nutrient solubilization, and increase in biomass and biocontrol (Mitter et al. 2013c; Sheng et al. 2008). These examples show that DEB_{OP} portray a wide range of endophytic lifestyles useful in microbe-assisted contaminant removal, especially in organopollutant-stressed environments.

5 Modes of Action of DEB_{OP} and Crops in Organopollutants Reduction: The Interplay Between Microbial Degradation and the "Green Liver" Concept

The degradative potential of DEB_{OP} is a determining feature in the detoxification of organic pollutants in plant niches. However, the beneficial endophytic bacteria-plant interactions may vary due to the physicochemical characteristics of the pollutant, crop physiological properties, and bacterial activity. After root uptake and translocation to other plant tissues, organopollutants may be subsequently metabolized, volatilized, or partially degraded, as part of their biotransformation into less phytotoxic compounds in tissue cells of crops (Fismes et al. 2002; Harms et al. 2004; Kvesitadze et al. 2001; Limmer and Burken 2016; Ma and Burken 2003; Pilon-Smits 2005; Pilon-Smits and Freeman 2006; Salt et al. 1998; Schröder 2007; Singh and Jain 2003). The phenomenon is referred to as the "green liver" concept, which depicts the fate and disposition of organic contaminants by plants (de Mello-Farias et al. 2011; Reichenauer and Germida 2008; Sandermann 1994, 1999; Stephenson and Black 2014). As diagrammatically represented in Fig. 1, plant-bacteria endophytes similarly have a part to play in in planta toxin degradation, with shared roles between the host plant and their associated bacteria. As a result of the presence of several toxic organic pollutants, this may affect the composition and role of the plant microbiome (the biomolecular network encompassing the host crop and its associated microbial communities) (Newman and Reynolds 2005). With pollutant-degrading traits in play during in planta detoxification (whether partial or full metabolism), this makes DEB_{OP} acclimatize to crops and the greater contaminated environments (Feng et al. 2017d; Kukla et al. 2014). Additionally, such plant-microbe interactions induce the co-metabolic degradation and metabolism of organics, which results in the dissipation of organopollutants in crops (Feng et al. 2017d; Pandotra et al. 2018; Tétard-Jones and Edwards 2016). The synergistic plant-bacteria partnerships are key to circumventing contamination of food crops grown in organopollutant-contaminated matrices (Khan et al. 2013a).

The metabolic performance of DEB_{OP} and their hosts are beneficial to the mutual partners, especially in crops. The manifestation of this shared detoxification strategy was seen by Sun et al. (2018), whereby they colonized the isolate *Pseudomonas* sp. Ph6-*gfp* in internal tissues of pak choi crops, which led to the biotransformation and prevention of phenanthrene. Their investigation demonstrated an interplay between plant metabolism and endophytic organopollutant degradation in crops, as analyzed using high-resolution mass spectrometry (HRMS) coupled with $^{13}C_2$-phenanthrene labeling (Sun et al. 2018). However, other studies point toward the microbial degradation of organopollutants in plant tissues, while plants facilitate residence to the microbial allies. The endophytic strain *Massilia* sp. Pn2 fostered plant growth in Italian ryegrass (*Lolium multiflorum* Lam.), even as phenanthrene levels in ryegrass roots and shoots reduced by 54% and 57%, respectively (Yousaf et al. 2011). Jabeen et al. ascribed the degradation of chlorpyrifos (CP) and its metabolite 3,5,6-trichloro-2-pyridinol to the efficient root colonization of *yfp*-tagged

Mezorhizobium sp. HN3, leading to a marked reduction of CP residues in roots and shoots of ryegrass (*Lolium multiflorum* var. Taurus) plants vegetated in CP-contaminated soils (Jabeen et al. 2016). Afzal et al. pointed out that the soil inoculation with *Burkholderia phytofirmans* PsJN was the most efficient method in improving plant biomass production, hydrocarbon degradation, and toxicity reduction, which correlated with the rhizo- and endospheric populations of the *Burkholderia* strain in the host plants (Afzal et al. 2013a). In a different study, toluene degradation and reduced evapotranspiration (by 30%) using the inoculate *Achromobacter xylosoxidans* F3B showed that the inoculated strain aided the vetiver grass (*Chrysopogon zizanioides*) plant in tolerating lethal concentrations of aromatic compounds while decreasing catechol- and phenol-induced phytotoxicity (Ho et al. 2013). These findings indicate that plants remain viable when microbial activity is enhanced, with some bacterial endophytes enabling their hosts to withstand high contaminant concentrations. However, further research is warranted in unlocking the hidden roles of both plants and their associated microorganisms during organopollutant decontamination in crops. Technologies such as HRMS coupled with isotope labeling and omics approaches are a step in the right direction in unveiling the full functions of the endosymbiont-host partnership.

The genomic characterization and gene expression of bacterial endophytes like DEB_{OP} have uncovered relevant information about the endophytic lifestyle and activity with other bacteria within their hosts (Mitter et al. 2013c; Piński and Hupert-Kocurek 2016). The genetic information (coding for pollutant-degrading enzymes) is a predetermined feature of pollutant-degrading endophytic bacteria for efficient organopollutant degradation and is located on mobile genetic elements such as bacterial plasmids or chromosomes (Wang et al. 2007, 2010). The genes encoding for enzymes involved in in planta organopollutant degradation coming from either plants or DEB_{OP} include cytochrome P_{450} enzymes, glutathione-*S*-transferase, peroxidase, polyphenol antioxidant enzyme, catalase, superoxide dismutase, esterase, amidase, hydrolase, and catechol-2,3-dioxygenase (C23O) (Alkorta and Garbisu 2001; Kvesidatza et al. 2009; Macek et al. 2000; Schäffner et al. 2002; Zhu et al. 2014). In addition to those genes, horizontal gene transfer (HGT) of pollutant-degrading and plant growth-promoting genes within the endospheric microbial population is a major mechanism by which microorganisms acquire new metabolic traits and rapidly adapt to organopollutant-stressed sites (Barac et al. 2004; Taghavi et al. 2005, 2009; van der Lelie et al. 2009; Weyens et al. 2010a, b). The manifestation of HGT (also called lateral gene transfer) was observed in the application of *Burkholderia cepacia* strain FX2 to important crop plants, corn (*Zea mays* L.), and wheat (*Triticum durum* Desf.), pointing to the horizontal transmission of C23O gene among the other endophytic bacteria (Wang et al. 2010). Consequently, the induced environmental remediation and plant growth-promoting attributes of the two plants by the endophytic *Burkholderia* strain, as well as disease suppression, were observed. The supposed acquisition of genes during lateral gene transfer may play a huge role in adapting the endogenous endophytic community to the organic-polluted environments: the degradation pathway is transferred among the members of the microbial community, as opposed to the integration of a new bacterium in a stable community (Newman and Reynolds 2005; van der Lelie et al. 2005). In herbicides, the genetic transfer (naturally or through recombinant DNA technology) of genes encoding for $BphK^{LB400}$ protein (bacterial glutathione-*S*-transferase) from *Burkholderia xenovorans* LB400 to endophytic bacteria provided their pea plant hosts with the degradative ability to remove 2,4-D, atrazine, aminotriazole, and pentanochlor (McGuinness et al. 2007). The microbe-assisted organopollutant removal approach has been investigated in commercially important tree species, such as *Populus trichocarpa* (Doty et al. 2017; Taghavi et al. 2005, 2009, 2010; Taghavi and van der Lelie 2013; van Aken et al. 2011; van der Lelie et al. 2009; Weyens et al. 2010b, 2012) and *Salix gooddingii* (Taghavi et al. 2009). However, there is a need for research on the HGT mechanism to confer PGP and pollutant-degrading qualities in bacterial endophytes in crop plants.

In summary, three particularly defining attributes of endophytic bacteria in in planta degradation of organic pollutants in crops have been reviewed (Afzal et al. 2014; Chandra and Singh 2016). These include (1) fostering plant growth and biomass production, (2) improving organopollutant bioavailability and innate degradative gene expression, and (3) enhancing population size and activity of indigenous bacteria to degrade POPs through horizontal gene transfer (HGT). As pointed out in Table 1, some of these postulates have been investigated in great detail in endophytes (as well as their endomicrobiome environment within plants, in the case of HGT) of various crops such as *Allium tuberosum* (Feng et al. 2017a), *Lolium multiflorum* (Yousaf et al. 2010a), *Lotus corniculatus* (Yousaf et al. 2010a), *Brassica chinensis* (Sun et al. 2014a, b, 2018), *Triticum aestivum* (Zhu et al. 2017a, b), and *Zea mays* (Wang et al. 2007, 2010). Finally, the interplay between DEB_{OP} and their hosts is seen in the bacterial influence in subcellular colonization of plant cells and plant enzymatic activities, whereby they impact the production of detoxification enzymes such as cytochromes P450 and glutathione-S-transferases (GST) or promote organopollutant-degrading gene expression in these plants (Feng et al. 2017a). These processes show that the indigenous and horizontally acquired degradative bacterial genes which encode for both PGP and pollutant removal traits are essential features in greatly enhancing crop endomicrobiomes (microbial communities in the shoot and root endospheres) which previously did not display these attributes while improving crop viability in organically contaminated niches.

6 Biotechnological Applications of DEB_{OP} in Crop Organopollutant Reduction

Since Galippe's discovery of bacterial endophytes in plants (Compant et al. 2012; Galippe 1887), DEB_{OP} research has made tremendous strides with various enhanced analytical methods, as shown in Fig. 2. However, this has translated into insufficient field applications of microbe-plant interactions involving DEB_{OP} and crops (Mercado-Blanco and Lugtenberg 2014; van Aken 2009). So far, research in

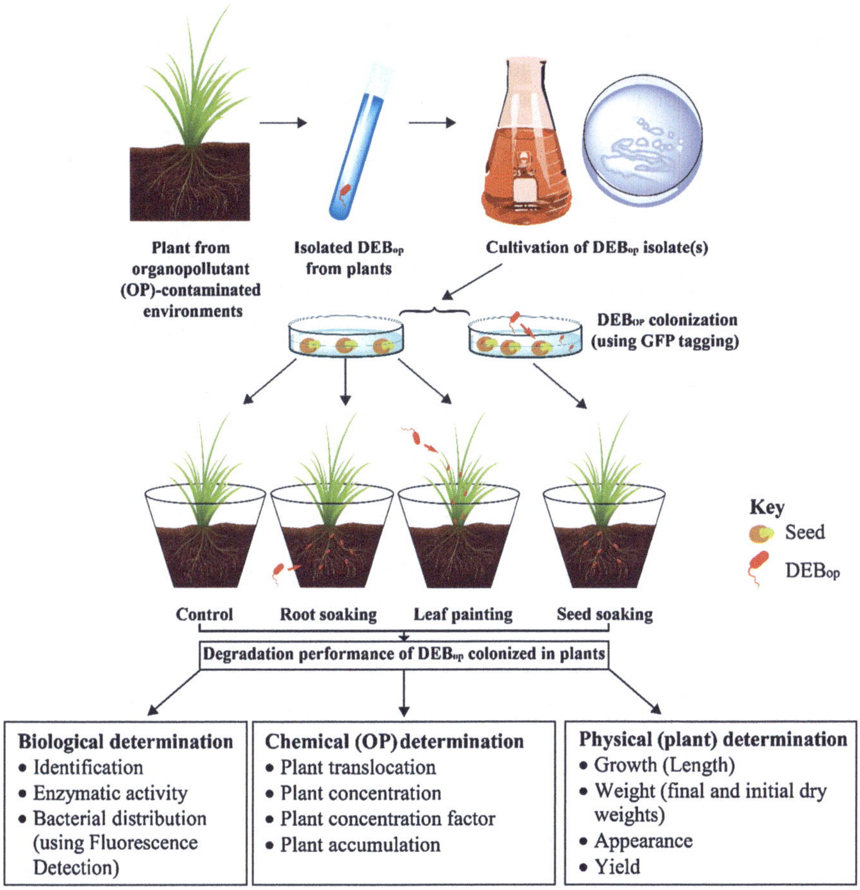

Fig. 2 Latest laboratory and greenhouse applications of DEB_{OP} in reducing crop POP contamination. The flow diagram explains the current process in the research and application of EB_{POP} in crop organopollutant detoxification. Aseptic isolation and cultivation may be followed up by techniques such as morphological (e.g., bacterial staining methods and colony formation and appearance), biochemical (such as enzymatic and nutrient reaction tests), and phenotypic (DNA-based methods). Also, colonization strategies (root/seed soaking or leaf painting) that induce mutualistic microbe-plant associations may be determined using analytical techniques of detecting the distribution and investigating the degradation performance of DEB_{OP}. These methods may be useful for bacterial studies, whether in isolation or inoculation with tested crop species in organic-contaminated soils in pot experiments, which represent the organopollutant-stressed environments

degradative endophytic bacteria has lagged significantly behind that of commensal, phytopathogenic, and PGP bacteria, in spite of the fact that numerous studies have shown that the health of the host depends on the right mix of commensal and mutualistic microbes degrading the organopollutants (Lacava and Azevedo 2013; Lodewyckx et al. 2002; Mastretta et al. 2006). This has necessitated the development of advanced biotechnological approaches that will elevate DEB_{OP} research and its applications in crop improvement and viability. The focus of the current and future applications of microbial technologies in screening for new strains and methods of selective inoculation will be expounded in the following subsections.

6.1 Omics Technologies in Crop Organopollutant Reduction

The fast-paced growth in advanced bacterial endophytic research has led to the realization of their adaptive roles in colonization, distribution, and mechanisms of bacterial endophytes, in terms of their lifestyle, phytopathology, and plant growth promotion in plants (Kaul et al. 2016). In contrast, a dearth of investigative studies exists in organopollutant-degrading bacterial endophytes and their roles in crops. This scenario has led to DNA-based methods such as next-generation sequencing, second-generation sequencing technologies (Loman et al. 2012a, b), and third-generation sequencing technologies (Bell et al. 2014; Goodwin et al. 2016; Shokralla et al. 2012). These technologies (abbreviated as NGS, SGS, and TGS) have ushered in the omics- and meta-omics era, with techniques such as metagenomics, metatranscriptomics, metaproteomics, metabolomics, mobilomics, lipidomics, and phenomics taking over from the preexisting genomic and proteomic methods (see Table 2 for their current and future uses). These DNA analytical techniques are utilized in deciphering the dual endophytic lifestyle of DEB_{OP} in adapting to various environmental pollutants and stimulating PGP mechanisms in plants (Barret et al. 2011). With the help of genomic databases such as GenBank, RefSeq, DNA Data Bank of Japan, and EMBL, these have paved the way for faster, low-cost acquisition and phenotypic analyses of genome sequences in a day (Thijs et al. 2016b). Owing to the benefits of utilizing omics technologies, this allows for extensive integration and interpretation of the omics information while offering a wide range of opportunities for optimization and further insight into microbe-assisted organopollutant removal (Nekrutenko and Taylor 2012).

The discovery of plants and bacterial endophytes, not only as autonomous entities but as "metaorganisms" (plant and its associated microorganisms, i.e., plant microbiota) (Thijs et al. 2016b), has deepened our understanding of their individual and shared roles, albeit with voluminous data (Howe et al. 2008). This revelation was observed in the following studies, whereby the use of these high-throughput sequencing technologies led to the uncovering of a broad variety of endophytic lifestyles in the bacterial strain *Burkholderia phytofirmans* PsJN in various interactions with plants, as shown in Table 3 (Coenye and Vandamme 2003; Mitter et al. 2013a). The knowledge disseminated may be useful in bioprospecting plant microbiomes for various biotechnological applications (Müller et al. 2016; Strobel and Daisy 2003). These include (1) microbe-based plant breeding for bio- and phytoremediation; (2) DEB_{OP} as biofertilizers to improve crop production in organic-contaminated and nutrient-stressed environments; (3) microbe-based plants as both phytoremediator and bioenergy crops; and (4) novel natural and synthetic

Table 2 "Omics" approaches in DEB_{OP} and crops in microbe-assisted organopollutant removal

Role of omics approach in DEB_{OP}-plant interaction	Technology (and examples)	Impact on knowledge	References
Identification of novel hydrocarbon-degrading strains isolated from crops	• SIP coupled with metagenomics (shotgun sequencing of SIP-derived metagenomes)	• Selective enrichment of targeted microbial populations, whose diversity is much less complicated than that of entire communities, with subsequent isolation of the particular functional metagenome of interest increases the feasibility of achieving coverage and assembly of individual genomes with significantly reduced efforts and sequencing cost	Thijs et al. (2018), Uhlik et al. (2013)
Deciphering the aspects of the endophytic lifestyle	• Metagenomic approaches (using 454 Genome sequencer FLX from 454 Life Sciences, Roche Applied Sciences, Indianapolis, IN)	• Useful in predicting traits and metabolic processes necessary for the endophytic lifestyle. This indicates that the endorhizosphere is an exclusive microhabitat requiring numerous adaptations	Sessitsch et al. (2011)
Understanding the effect of RECs on biodegradation root exudates	• SIP labeling of RECs	• SIP labeling may be employed to elucidate the role of RECs in hydrocarbon degradation	Thijs et al. (2018), Uhlik et al. (2013)
Plant-microbe interactions	*Omics techniques* • Lipidomics • Metabolic engineering and systems biology tools • Multi-omics approaches	• Lipidomic studies may be useful in not only identifying additional lipid molecules involved in plant-microbe interactions but also revealing lipid patterns that are common or different among various interactions between plants and microbes (i.e., phytopathogenic, commensal, or beneficial) • General understanding plant-microbe interactions and to improve plant traits • Useful in understanding the abiotic	Siebers et al. (2016), Vishal et al. (2016), Meena et al. (2017)

(continued)

Table 2 (continued)

Role of omics approach in DEB$_{OP}$-plant interaction	Technology (and examples)	Impact on knowledge	References
		stress responses in plants and microbe-mediated stress mitigation	
Metaorganism (the aspect of a plant as a biomolecular network composed of the host and its associated endosymbiont/ microbiota)	• Environmental DNA shotgun metagenomics • High-throughput metatranscriptomics • Powerful imaging techniques	• Important in the selection and management of complementary plants and microbiomes • Useful in the study of the interactions between hyperaccumulator plants and their microbial rhizobiome on the whole plant level	Thijs et al. (2018)
Relationships between microbial identity and particular metabolic functions	• DNA-SIP coupled with metagenomics	• Allows the detection of rare low-abundance species from metagenomic libraries • Facilitates the detection of novel enzymes and bioactive compounds	Chen and Murrell (2010)
Microbial taxonomy composition to predict hydrocarbon degradation rate?	• Next-generation sequencing	• Establishing predictive relationships between microbial community structure and biotransformation rates, which facilitates monitoring and follow-up of shifting microbial activities over time and allows optimizing phytoremediation as contaminant composition changes • Revealing the shifts in the chemical composition of diesel constituents and the abundance of specific OTUs, as investigated by pyrosequencing, suggested that natural attenuation has occurred at a diesel-contaminated railway site	Chikere et al. (2011), Sutton et al. (2013)
Mining for genes that determine endophytic life Species diversity changes due to	• Comparative genomic analyses • Comparative genomic analyses using Ion Torrent Personal	• Identification of various genetic factors that determine endophytic colonization of plants and sustaining bacterial	Ali et al. (2014), Mitter et al. (2013a, b, c), Tardif et al. (2016)

(continued)

Table 2 (continued)

Role of omics approach in DEB_{OP}-plant interaction	Technology (and examples)	Impact on knowledge	References
increased contaminant levels	Genome Machine (PGM)	life within the host plant. This has implications for inoculant development • Increasing PHC contamination levels resulted in shifts in the microbiome composition, favoring putative hydrocarbon degraders, and microbes previously reported as associated with plant health of two willow cultivars (*Salix purpurea* cv. Fish Creek and *Salix miyabeana* cv. SX67)	
Mining for the presence of MGEs (such as plasmids with catabolic genes)	• Mobilomics research in the analysis of bacterial MGEs in complex environments (also known as mobilomics/metabolomics/plasmid metagenomics/plasmidomics)	• Metagenomic mining of the "mobilome" (i.e., the preferential extraction of plasmid DNA) to access the ongoing horizontal gene transfer	Jørgensen et al. (2015)
Impact of endophytic microbe-assisted phytoremediation in bioenergy crops	• Metabolomics and proteomics	• Evaluating the effect of contaminant-degrading bacteria inoculation on organopollutant uptake by maize plants grown in pollutant-stressed soils	Li et al. (2014)

Recent and futuristic applications of multi-omics approaches in revealing the beneficial microbe-plant interactions between DEB_{OP} and crops in in planta organopollutant removal. *DNA-SIP* DNA-stable-isotope probing, *MGEs* mobile genetic elements, *OTUs* operational taxonomic units, *RECs* root exudate compounds, and *SIP* stable-isotope probing

strains and bioresources for improved in situ or ex situ degradation, e.g., bioresources such as microbes, enzymes, and secondary metabolites (Singer et al. 2003; Thijs et al. 2018; Weyens et al. 2009b). The novel insights made by the biocuration of the multidisciplinary data obtained from the abovementioned state-of-the-art tools will tremendously increase our understanding and approach of the mutualistic plant-microbe interactions in the natural crop context (Krishnaraj and Pasha 2017).

Table 3 DEB_{OP} genomes influencing organopollutant detoxification and plant growth promotion in plants

Bacteria	Genomic size (replicons)	Isolated plant body	References
Paraburkholderia phytofirmans PsJN (*Burkholderia phytofirmans* PsJN^T)	8.2 Mb genome (2 circular chromosomes and 1 plasmid with a total of 7,405 genes)	*Glomus vesiculiferum*-infected onion roots	Weilharter et al. (2011)
Methylobacterium radiotolerans	6.8 Mb draft genome (1 circular chromosome with 6,373 coding genes)	Zucchini plant (*Cucurbita pepo* L.)	Eevers et al. (2015a, b, c)
Sphingomonas taxi	3.9 Mb draft genome (1 circular chromosome with 3,634 coding genes)		
Enterobacter aerogenes	4.5 Mb draft genome (1 circular chromosome with 4,191 coding genes)		
Serratia proteamaculans 568	5.5 Mb (1 chromosome and 1 plasmid)	Soybean	Taghavi et al. (2009)

Complete and draft genomes useful as starting points for future "omics" research

6.2 Direct Application of Selectively Compatible Consortial Strains

The enhancement of organopollutant detoxification is imperative for crop sustainability. In planta decontamination can be improved through the direct application of selectively compatible strains (Gaiero et al. 2013; Rylott 2014). This method creates an optimal microenvironment for organopollutant removal in crops, a critical factor during the colonization and distribution of these microbes in the microbe-assisted removal of pollutant organics. A consortium of various microorganisms, combined to form a composite of endophytic bacteria, perform better than individual microorganisms (Gkorezis et al. 2014, 2016; Pawlik et al. 2017; Wang et al. 2017a). However, interspecies competition between these microorganisms may lead to suboptimal population ratios and difficulty in preserving the optimal spatial distribution, leading to the reduction in remediation efficacy of organopollutants and incomplete detoxification (Tripathi et al. 2017). The selection of compatible microbial strains and the provision of an optimal microenvironment in crops for endospheric colonization are crucial in the development of efficient consortial endosymbionts (Thijs et al. 2016a). Enabling approaches such as omics strategies would assist in analyzing the cohesion of microorganisms and plants via their -omics and meta-omics data. This will aid in the selection of the fitting endophytes in the plant microbiome environment essential to the in planta organopollutant detoxification.

6.3 *Exploitation of DEB_{OP} in Tripartite Associations with Fungi and Plants*

Since the evolution of plants inhabiting the earth, bacterial symbionts have been prevalent in various terrestrial niches, with the coevolution of such microbes to form tripartite associations with mutualistic fungi and plants (Badri et al. 2009; Martin et al. 2017). The microbial members of this ancestral alliance facilitate a role in either PGP or organopollutant reduction in crops, while the plants provide nourishment and residence to its microbiota (Alarcón et al. 2008; Waigi et al. 2017; Jambon et al. 2018; Martin et al. 2017). This occurs due to the modified composition of the rhizospheric microbial community structure, depending on the nature of the root exudates and the type of arbuscular mycorrhizal fungi (AMF) associated with a crop, before the selective establishment of the telluric microflora into the crop endosphere. As AMF contribute to plant growth and root organopollutant uptake by crops (Gao et al. 2010), they also create favorable microenvironments (mycorrhizosphere/hyphosphere effect) that intensify the activity of pollutant degraders such as DEB_{OP}. The bacterial endophytes take up the role of mycorrhizal helper bacteria, in their coexistence and organopollutant detoxification within the host crop (Alarcón et al. 2008; Waigi et al. 2017). This shows that the application of hydrocarbonoclastic/organopollutant-degrading bacterial endosymbionts capable of stimulating PGP in their hosts may lead to positive interaction effects. A more detailed study on the types and examples of these complex plant-bacteria-fungi interactions has been reviewed by Jambon et al. (2018). The connections between fungi and bacteria are of particular interest in applied techniques, as crops take advantage of the attributes of the mutualistic microorganisms (i.e., improving crop production and organopollutant detoxification).

6.4 *Transgenic Applications of DEB_{OP}*

Though plants have several advantages over bacteria in organopollutant bioremediation (environmentally friendly, widely socially acceptable, sustainable, and easy to harvest the sequestered contaminants), they are unable to mineralize numerous complex xenobiotics on their own (Weyens et al. 2009c). Their limited ability to detoxify and compartmentalize these compounds either in lignified cell walls (apoplast) or in vacuoles leads to their bioaccumulation within cells or cell envelopes in crops (Morsy 2015). Consequently, studies conducted on endophytic bacteria with better degradation qualities have been on the rise (Barac et al. 2004; Taghavi et al. 2005, 2009), as these chemicals are poisonous in their parent or intermediate states in plants. Genetic engineering and transgenics, therefore, have accelerated in bridging the gap between plants and engineered endophytes or their catabolic genes (Abhilash et al. 2009; van Aken et al. 2010; de Mello-Farias et al. 2011).

Transgenic plants (which express bacterial or mammalian genes involved in metabolism, uptake, or transport of specific organopollutants) or their associated DEB_{OP} which degrade trichloroethylene (TCE), pentachlorophenol (PCP), and polychlorinated biphenyls (PCBs), among others, have already been documented (Abhilash et al. 2009; Barac et al. 2004; de Mello-Farias et al. 2011; Kawahigashi 2009; van Aken 2008, 2009, 2011; van Aken et al. 2010). An understanding of the strategy involved in genetic engineering and transgenic technology has been documented in the cited reviews and studies (Cherian and Oliveira 2005; Dietz and Schnoor 2001; Doty 2008; Doty et al. 2000; French et al. 1999; Macek et al. 2008; Rosser et al. 2001; van Aken 2009; van Aken et al. 2010). These technologies are crucial in overcoming the inherent drawbacks of plant detoxification capabilities (Doty et al. 2007), much like the strategy similar to the development of the genetically modified crops through *Agrobacterium*-mediated plant transformation (Eapen et al. 2007; Gelvin 2003; Van Aken 2009). However, there is a lack of evidence-based documentation of employing transgenic bacteria or plants in the organopollutant reduction in crops. Furthermore, transgenics in various phytoremediation approaches have only been applied to bioenergy crops and trees in various polluted environments, a useful approach in utilizing such plants to rehabilitate organopollutant environments (Witters et al. 2012a, b). On the other hand, concerns from field testing, environmental risk assessment, and legal issues have to be addressed before transgenic plants or microbes are commercially utilized for in planta xenobiotic pollutants removal (de Mello-Farias et al. 2011; Ryan et al. 2008).

6.5 DEB_{OP} in Biofertilization

Beneficial bacteria have been exploited as inoculants and can be classified as biofertilizers (such as *Rhizobium* spp.), phytostimulators (such as *Azospirillum*, a group of auxin-producing, root-elongating bacteria), biopesticides, and bioremediators (pollutant degraders which use carbon sources such as root exudates in the case of rhizoremediators or organopollutants in the case of both DEB_{OP} and rhizoremediators) (Berg 2009; Bloemberg and Lugtenberg 2001; Kuiper et al. 2001, 2002; Lugtenberg et al. 2002). Biofertilizers are referred to as products based on microorganisms that improve the nutritional status crops (James 2000). Biofertilizer application by beneficial bacteria accounts for approximately 65% of the nitrogen supply of crops worldwide (Lugtenberg et al. 2002). In this sense, rhizobial inoculants in legumes that fix atmospheric nitrogen in root nodules are considered to be "first-generation biofertilizers" (Mulas et al. 2013). Furthermore, endophytic inoculant strains living inside the root tissues and undergoing nitrogen fixation are collectively known as "second-generation biofertilizers" (Mulas et al. 2013). As such, the potential exploitation of the dual endophytic lifestyle of DEB_{OP} may be crucial in enhancing a crop's viability in organic-contaminated environments.

As bacteria have been exploited for their PGP attributes for the past four decades, their utilization as organopollutant degraders has only been well known recently. For example, the use of endophytic bacterial components that play a role in the recognition of the plant in various stages of the nodulation process have been identified, including the Nod factors, extracellular polysaccharides (EPSs), lipopolysaccharides (LPSs), and periplasmic cyclic glucans (Lugtenberg et al. 2002). Nitrogen, a vital macronutrient in crop survival, and its fixation have been a critical area of interest in endophytic bacterial studies, with diazotrophs from taxa such as *Rhizobium* sp. in leguminous crops (Oldroyd et al. 2011), *Frankia* in woody species (Huss-Danell 1997), as well as *Azospirillum*, *Bacillus*, and *Enterobacter* in nonleguminous plants (Bahulikar et al. 2014; Puri et al. 2017). However, a research area to look into for beneficial bacteria such as DEB_{OP} isolates is on biological nitrogen fixation, which would be an essential asset for pollutant-degrading endophytes to adapt to pollution in the ever-changing global environment (Afzal et al. 2014; Imadi et al. 2015). The exploitation of a suite of the cutting-edge technologies to identify the most effective endophytic diazotrophs may be viable alternatives to chemical fertilizers in combating N-deficiency in various terrestrial niches (Doty et al. 2012, 2017a). Additionally, the use of diazotrophic bacterial endophytes may result in sustainable agriculture, bioenergy crop production, and forestry more environmentally and economically (Doty 2017b). This will lead to further research into bacterial endophyte-assisted crop production, thus providing the basis for a better understanding of the multiple interactions taking place in the plant environment (Sessitsch et al. 2016), especially in organopollutant detoxification and agricultural sustainability.

7 Advantages of Using DEB_{OP} in Crop Organopollutant Decontamination

Crops and their associated microorganisms live in a symbiotic relationship that aids them in surviving organic-contaminated environments. The presence of beneficial microbiota such as DEB_{OP} in the plant metaorganism facilitates a more efficient degradation of organic compounds they encounter in crops (Eevers et al. 2017). As plants take up organopollutants, their interplay with bacterial endosymbionts facilitates the breakdown of the toxic parent compounds to nonharmful end products. This cooperation assists plants in adapting to organopolluted niches, as DEB_{OP} get attuned to the excess carbon from the phytouptake of toxins. Below are the advantages of DEB_{OP}, in terms of their comparisons with both rhizobacteria and crops, as well as their application during in planta organopollutant detoxification.

Following phytouptake, toxic compounds undergo metabolism and/or evapotranspiration (the release of parent compounds or their metabolites into the atmosphere via aerial plant parts). Though plants usually metabolize or compartmentalize xenobiotics, they are unfavorable compared to DEB_{OP} involvement in crop organopollutant detoxification in the following ways (Burken et al. 2000;

Burken 2004): (1) plants have not evolved and undergone selective pressure necessary in acquiring the degradative capacity to metabolize various recalcitrant pollutant organics. This is due to the photoautotrophic nature of crops, as they are not dependent on organic molecules as a carbon or energy source. (2) Crops may convert organopollutants to their conjugated forms through the green liver concept, which avoids phytoaccumulation and potential phytotoxicity of organic toxins. However, plant metabolism alone is a limited approach, as cell organelles may accumulate the conjugated toxins and/or intermediate metabolites. In contrast, DEB_{OP} during in planta organopollutant degradation have undergone evolution and selective pressure, as well as organopollutant mineralization terminating with the conversion into CO_2, water and cellular biomass (Weyens et al. 2009c).

Regarding the degradation of xenobiotic compounds, endophytes may have several advantages as compared to rhizosphere bacteria. Even though endophytic and rhizospheric populations are conditioned by biotic and abiotic factors, endophytic bacteria are well protected from these stresses than their rhizospheric relatives (Afzal et al. 2011; Rosenblueth and Martínez-Romero 2006; Seghers et al. 2004; Segura and Ramos 2013). The protection benefit by the host assists the bacterial endophytes in establishing within the plant tissues, making DEB_{OP} to not compete with soil bacteria for binding sites on host plant root surfaces and nutrients (Chaturvedi et al. 2016; Ryan et al. 2008). In addition, endophytic isolates are capable of colonizing or recolonizing internal plant tissues in higher numbers than the isolates from the root surface which shows how they adaptively benefit their host partners (Afzal et al. 2011; Andria et al. 2009). Moreover, bacterial endophytes have a significant role in metabolizing organopollutants, as pollutant-degrading bacteria are more abundant among endophytic populations than in the rhizosphere of plants in contaminated sites (Rosenblueth and Martínez-Romero 2006; Siciliano et al. 2001). Finally, an important advantage is that DEB_{OP} are efficient rhizospheric and endospheric colonizers and may degrade contaminants in both environments.

The application benefits of DEB_{OP} in microbe-assisted organopollutant removal in crops is vital to the concerted efforts in detoxification between the mutual partners. Endophytic bacteria are easier to manipulate than plants in cases where the genetic transfer of a xenobiotic degradation pathway is required (Ryan et al. 2008). Moreover, the quantitative gene expression of organopollutant degradative genes within the endophytic populations is a useful monitoring tool for evaluating the efficacy of the remediation process (Khan and Doty 2011). Also, DEB_{OP} may even interact with the plant tissues in the efficient synergistic degradation of contaminants in the plant environment (Afzal et al. 2014; Glick 2010; Kandel et al. 2017; Khan et al. 2013b; Weyens et al. 2009b). Another significant advantage of using endophytic pollutant degraders is that they reduce crop phytotoxicity through the in planta degradation of xenobiotics taken up by the plant (Afzal et al. 2014; Eevers et al. 2017; Ryan et al. 2008). The overall success of employing endophytic bacteria-plant associations in crops organopollutant reduction is only achieved when all factors and features detailed above are capitalized on for crop viability.

8 Challenges in Exploiting DEB_{OP} in Reducing Crop Organopollutants Contamination

As seen in the previous sections, organopollutant-degrading endophytic bacteria are well adapted with various attributes, be it in their colonization and distribution, mechanisms of survival, and degradation performance in crop detoxification. Furthermore, the current and potential applications of DEB_{OP} that may biotechnologically improve plant growth and adaptability in organopollutant-contaminated environments have been mentioned. However, certain disadvantages arise during their deployment in reducing the impact of organopollutants in crop production (Khan and Doty 2011). The limitations detailed below are divided into crop, DEB_{OP}, biotic, and abiotic factors, as well as implementing factors of this remedial strategy.

Crop constraints that are absent in the laboratory and greenhouse studies may result in considerable limitations in field applications (Gerhardt et al. 2009, 2017), hence having a significant impact on endophytic bacterial activity (Hallmann and Berg 2006). For example, the seasonality of a choice of crop to be applied in in planta organopollutant detoxification shows its dependence on the season and the length of time in growth (Beans 2017). Furthermore, the designated vegetated species in this remedial strategy should be accommodative to its microbial flora (Weyens et al. 2009c). Otherwise, it may aid in the conversion of the beneficial endophytic bacteria to become pathogenic or commensalistic microbes (Brader et al. 2017), which may be disadvantageous to organopollutant detoxification. Therefore, the crops to be planted should be adaptive to its host role for its endosymbiotic microbiota and the environment in order to facilitate in planta decontamination.

The selection of endophytic bacteria to be used in microbe-assisted removal of organopollutants in crops should take into account the following shortcomings. Some DEB_{OP} may convert to become either phytopathogenic or inactive strains, due to HGT from other endophytic bacteria in the microbial community (Brader et al. 2017). This transformation of DEB_{OP} may be due to higher concentrations of the organopollutants (Barac et al. 2004; Taghavi et al. 2005) or the impact of the surrounding endophytic microbial communities (Berg et al. 2008). As a result of the transferred genes from surrounding microorganisms, the biotransformation of volatile, highly water-soluble organic xenobiotic compounds such as benzene, toluene, ethylbenzene, and xylene is often limited due to insufficient degradation of the pollutants by the host crop or the DEB_{OP} (Montpetit and Lachapelle 2017). This incomplete detoxification results in the phytotoxicity of these compounds and their metabolites (such as reactive oxygen series), which potentially gives rise to new environmental problems (Liu et al. 2017; Schwitzguébel 2017). Though the use of microbial inoculants like DEB_{OP} as biofertilizers may establish very high population densities immediately after inoculation, their densities may decline over time and distance from the inoculum source. A further drawback is the potential risks associated with their release in whichever application approach, with unknown effects over native microbial communities and the wider environment.

In terms of abiotic factors, a downside to the in planta organopollutant detoxification treatment includes the fact that a specific organopollutant decontamination "prescription" cannot be applied to every site with a particular organic contaminant because different environmental conditions (e.g., soil and climate) may not be suitable for the target crop (Schwitzguébel 2017). For example, the environmental and climatic factors may greatly influence the entry of DEB_{OP} in their colonization and distribution in crops. Besides, in many field situations, in planta organopollutant removal may be applied to co-contaminated sites, which is a complicated issue in remediating co-contaminated areas. Heavy metals have a significant impact on the biodegradation of the taken up organopollutants by influencing the physiology and ecology of DEB_{OP} (Weyens et al. 2010b). Additionally, any bioremediation strategy involving endophytic bacteria is restricted to sites where the concentration of the organopollutant is not toxic to the proposed plant for decontamination. An important constraint to consider is the solubility of the pollutant that may hamper the ability of the plant to expedite organopollutant phytouptake (Dhankher et al. 2012). As a result, microbe-assisted organopollutant removal using live plants is restricted to sites conducive to the growth of the selected plant with the contaminant located within the potential root zone of the selected plant (Susarla et al. 2002).

With regard to the drawbacks in implementing in planta organopollutant removal, the social and regulatory matters surrounding bacterial use in the environment have been a thorny issue worldwide (Gerhardt et al. 2009, 2017). The sociopolitical context, which entrenches the introduction of microbe-assisted removal technologies using endophytes, is unique to each community and determines which technologies are either controversial or socially acceptable (Weir and Doty 2016). A case in point is the utilization of transgenics in the environment, which is a hot topic in plant biotechnology due to the legal and environmental issues surrounding genetically modified living organisms (Eapen et al. 2007). Biofertilizer application using endophytic bacteria is an approach which should be fully understood before applying into important crops, despite their beneficial advantages in nutrient availability and PGP attributes (Gaiero et al. 2013). Given that both techniques above are generally aesthetically pleasing, the acceptability of endophyte-assisted crop detoxification technology has the potential to be quite high. However, the association of food crops with such technologies is still a global issue currently due to food and environmental safety, and care should be taken in safety assessment and approval before implementing them (Sessitsch et al. 2019). A further bottleneck is that the present methods of assessing phytoremediation are not accurate enough to show decreasing levels of the remediated contaminant, even though active remediation may occur in several cases (Gerhardt et al. 2009, 2015).

As mentioned earlier, the disadvantages of in planta organopollutant removal in crops potentially hinder these allies in survival in organopolluted ecological niches. The controlled implementation of the detoxification strategy is vital to its success, with recommendations being enumerated below:

1. For inoculation of pollutant-degrading endophytic microbes into native crops, the careful selection of beneficial bacterial strains is key to avoiding some of the constraints related to the phytotoxicity and plant survival in non-native locations.
2. In the case of selecting indigenous bacteria for microbe-assisted removal, the strains involved should not turn to become pathogenic or neutral inside its host.
3. Research into both the crops and the associated microorganisms should be intensified in order for a breakthrough to be observed in in planta detoxification strategy. Whole genome sequencing should be stepped up in order to fully grasp the endophytic lifestyle of DEB_{OP}, as well as their host genes. PGP traits useful in crops like nitrogen fixation should be investigated in DEB_{OP} microorganisms. The use of omics and meta-omics approaches would lead to a massive step forward in endophytic bacterial research in crops grown in organopollutant-stressed niches.
4. The crop choice to be planted in in planta organopollutant decontamination must be versatile to its endosymbionts, the harsh environment, and its seasonality for its survival. It should have plant-associated genes required for metabolism and can take up various consortial microorganisms useful in detoxification.
5. The controlled use of transgenic technologies and other microbial techniques is critical in ensuring societal approval and environmental application. The recombined and horizontally acquired genes should not make the plants and microbes be detrimental for human and environmental health. It is, thus, recommended for bioenergy production or reforestation of organopolluted environments so that they have a direct nonfood-related benefit to both man and the environment (Ryan et al. 2008).
6. The advent of new techniques such as stable isotope probing (SIP), compound-specific isotope analysis (CSIA), HRMS coupled with isotopic labeling, Ion Torrent Personal Genome Machine (PGM), and various cutting-edge microscopes among others may be useful in illuminating the impact of organopollutant degradation by DEB_{OP} within crops. This will have far-reaching outcomes in the exploitation of the endosymbiotic lifestyle of DEB_{OP} in their hosts.

9 Conclusions

Microbe-assisted organopollutant removal has a great potential in reducing in planta contamination, while offering an aesthetic and economically efficient remedial approach, compared to physicochemical strategies. Additionally, microbe-assisted organopollutant removal can boost the long-term target of contaminant-free arable agricultural productivity, especially in pollutant-stressed environments. Though there is notable progress made in uncovering the colonization and distribution aspects of DEB_{OP} in crops, a significant information gap arises in the interplay pathways associated with bacteria-plant partnerships. The steady development of high-throughput analytical techniques in bacteria and plant genomic revolution has enabled the understanding of the metaorganism, which comprises the plant and its underlying microbiome. This revelation has fast-tracked the discovery and knowledge of biotransformation and biodegradation pathways displayed by the bacterial endophytes (PGP and pollutant degradation) in organopollutant-stressed farms. Advanced sequencing technologies will stimulate the interest in illuminating the genetics behind bipartite (bacteria-plant) and tripartite (bacteria-fungi-plant) associations exhibited by the metaorganism. Furthermore, established analytical tools used to study the environmental fate of different contaminants will ensure the traceability of the transport and degradation of environmental organopollutants, especially in the areas of qualitative and quantitative assessments of these toxins.

As DEB_{OP} research unveils new insights into their various aspects in microbe-assisted organopollutant removal in crops, this has led to challenges regarding their laboratory and field applications. As recent studies have revealed the tremendous progress made in *Burkholderia* research, the advanced omics approaches revealed vital information in the progress of endophytic bacterial research. However, much work is still required in advancing the knowledge on bacterial endophytic research in crops, especially in omics research in in planta organopollutant detoxification. Furthermore, bottlenecks such as biotic, abiotic, and pollutant factors may hamper the advancement of DEB_{OP} applications in crop organopollutant reduction into the field. Also, inoculant strains are often rapidly out-competed by the natural microflora that are generally better adapted to the environment. An additional challenge of applying transgenics technology in crops or bacteria is the legal and social implications surrounding it. In sum, microbe-assisted organopollutant removal in crops is a vital detoxification strategy utilizing indigenous or inoculated bacteria with known pollutant-degrading and PGP traits; however, their application mainly depends on the drawbacks faced in their use. Thus, the solutions to these limitations would enhance microbe-assisted organopollutant detoxification as a viable, attractive, green, and cost-effective alternative for the treatment of crops contaminated by organic pollutants. Additional research prospects could further explore the sociopolitical acceptability of microbe-assisted organopollutant removal strategy in diverse environmental settings and for various types of organic contaminants to ascertain the implementation potential of the technology.

10 Summary

Global contamination through organopollutant use is a modern-day environmental problem, especially in farmlands. The use of mutualistic bacteria-plant interactions in the in planta degradation of organic contaminants is an advocated strategy, as dealing with the availability of these toxins in soils is a major issue currently. An effective, environmentally friendly approach utilizes nonpathogenic colonizing bacteria that facilitate plant growth promotion and organopollutant degradation within plant tissues. The critical examination of the aspects of colonization, distribution, plant growth promotion, and degradative performance of the decontamination method were reviewed herein. The investigated literature data provides clear evidence of the viability of this technique in organopollutant degradation in polluted environments, especially when employing bioenergy crops and innate or inoculated bacteria. Though challenges may occur during the implementation of in planta crop detoxification, the advantages and the advanced analytical technologies offered will improve on the toxicological reduction in man's most well-managed resource. Policy measures regarding the use of bacteria in crops should be more emphasized with monitoring studies to combat the negative impact of organopollutants worldwide.

Acknowledgments This work was financially supported by the National Key Research and Development Program of China (2016YFD0800203), the National Natural Science Foundation of China (41877125, 41771523), and the Jiangsu Provincial Key Research and Development Plan, China (BE2017718). The authors are also appreciative of the valuable suggestions that the editor and the anonymous reviewers made in improving the quality of this manuscript.

Conflict of Interest The authors declare no competing interests.

References

Abhilash PC, Jamil S, Singh N (2009) Transgenic plants for enhanced biodegradation and phytoremediation of organic xenobiotics. Biotechnol Adv 27:474–488. https://doi.org/10.1016/j.biotechadv.2009.04.002

Afzal M, Yousaf S, Reichenauer TG, Kuffner M, Sessitsch A (2011) Soil type affects plant colonization, activity and catabolic gene expression of inoculated bacterial strains during phytoremediation of diesel. J Hazard Mater 186:1568–1575. https://doi.org/10.1016/j.jhazmat.2010.12.040

Afzal M, Yousaf S, Reichenauer TG, Sessitsch A (2012) The inoculation method affects colonization and performance of bacterial inoculant strains in the phytoremediation of soil contaminated with diesel oil. Int J Phytoremediation 14(1):35–47. https://doi.org/10.1080/15226514.2011.552928

Afzal M, Yousaf S, Reichenauer TG, Sessitsch A (2013a) Ecology of alkane-degrading bacteria and their interaction with the plant. In: de Bruijn FJ (ed) Molecular microbial ecology of the rhizosphere. Wiley, Hoboken, pp 975–989. https://doi.org/10.1002/9781118297674.ch92

Afzal M, Khan S, Iqbal S, Mirza MS, Khan QM (2013b) Inoculation method affects colonization and activity of *Burkholderia phytofirmans* PsJN during phytoremediation of diesel-contaminated soil. Int Biodeter Biodegr 85:331–336. https://doi.org/10.1016/j.ibiod.2013.08.022

Afzal M, Khan QM, Sessitsch A (2014) Endophytic bacteria: prospects and applications for the phytoremediation of organic pollutants. Chemosphere 117:232–242. https://doi.org/10.1016/j.chemosphere.2014.06.078

Ahemad M, Kibret M (2014) Mechanisms and applications of plant growth promoting rhizobacteria: current perspective. J King Saud Univ-Sci 26:1–20. https://doi.org/10.1016/j.jksus.2013.05.001

Alarcón A, Davies FT, Autenrieth RL, Zuberer DA (2008) Arbuscular mycorrhiza and petroleum-degrading microorganisms enhance phytoremediation of petroleum-contaminated soil. Int J Phytoremediation 10:251–263. https://doi.org/10.1080/15226510802096002

Ali N, Sorkhoh N, Salamah S, Eliyas M, Radwan S (2012) The potential of epiphytic hydrocarbon-utilizing bacteria on legume leaves for attenuation of atmospheric hydrocarbon pollutants. J Environ Manage 93:113–120. https://doi.org/10.1016/j.jenvman.2011.08.014

Ali S, Jin D, Charles TC, Glick BR (2014) A bioinformatics approach to the determination of genes involved in endophytic behavior in *Burkholderia* spp. J Theor Biol 343:193–198. https://doi.org/10.1016/j.jtbi.2013.10.007

Ali S, Charles TC, Glick BR (2017) Endophytic phytohormones and their role in plant growth promotion. In: Doty SL (ed) Functional importance of the plant microbiome: implications for agriculture, forestry and bioenergy. Springer, Cham, pp 89–105. https://doi.org/10.1007/978-3-319-65897-1_6

Alkorta I, Garbisu C (2001) Phytoremediation of organic contaminants in soils. Bioresour Technol 79:273–276. https://doi.org/10.1016/S0960-8524(01)00016-5

Andria V, Reichenauer TG, Sessitsch A (2009) Expression of alkane monooxygenase (*alkB*) genes by plant-associated bacteria in the rhizosphere and endosphere of Italian ryegrass (*Lolium multiflorum* L.) grown in diesel contaminated soil. Environ Pollut 157:3347–3350. https://doi.org/j.envpol.2009.08.023

Arslan M, Imran A, Khan QM, Afzal M (2017) Plant–bacteria partnerships for the remediation of persistent organic pollutants. Environ Sci Pollut Res 24:4322–4336. https://doi.org/10.1007/s11356-015-4935-3

Ashraf MA (2017) Persistent organic pollutants (POPs): a global issue, a global challenge. Environ Sci Pollut Res 24:4223–4227. https://doi.org/10.1007/s11356-015-5225-9

Bacon CW, White J (2000) An overview of endophytic microbes: endophytism defined. In: Bacon CW, White J (eds) Microbial endophytes. Marcel Dekker Inc, New York, pp 3–30

Badri DV, Weir TL, van der Lelie D, Vivanco JM (2009) Rhizosphere chemical dialogues: plant–microbe interactions. Curr Opin Biotechnol 20:642–650. https://doi.org/10.1016/j.copbio.2009.09.014

Bahulikar RA, Torres-Jerez I, Worley E, Craven K, Udvardi MK (2014) Diversity of nitrogen-fixing bacteria associated with switchgrass in the native tallgrass prairie of northern Oklahoma. Appl Environ Microbiol 80:5636–5643. https://doi.org/10.1128/AEM.02091-1

Balseiro-Romero M, Gkorezis P, Kidd PS, Van Hamme J, Weyens N, Monterroso C, Vangronsveld J (2017) Use of plant growth promoting bacterial strains to improve Cytisus striatus and Lupinus luteus development for potential application in phytoremediation. Sci Total Environ 581–582:676–688. https://doi.org/10.1016/j.scitotenv.2016.12.180

Barac T, Taghavi S, Borremans B, Provoost A, Oeyen L, Colpaert JV, Vangronsveld J, van der Lelie D (2004) Engineered endophytic bacteria improve phytoremediation of water-soluble, volatile, organic pollutants. Nat Biotechnol 22:583. https://doi.org/10.1038/nbt960

Barret M, Morrissey JP, O'Gara FJB (2011) Functional genomics analysis of plant growth-promoting rhizobacterial traits involved in rhizosphere competence. Biol Fertil Soils 47:729–743. https://doi.org/s00374-011-0605-x

Beans C (2017) Core concept: phytoremediation advances in the lab but lags in the field. Proc Natl Acad Sci U S A 114:7475–7477. https://doi.org/10.1073/pnas.1707883114

Becerra-Castro C, Kidd PS, Prieto-Fernández Á, Weyens N, Acea M-J, Vangronsveld J (2011) Endophytic and rhizoplane bacteria associated with *Cytisus striatus* growing on hexachlorocyclohexane-contaminated soil: isolation and characterisation. Plant and Soil 340:413–433. https://doi.org/10.1007/s11104-010-0613-x

Becerra-Castro C, Kidd PS, Rodríguez-Garrido B, Monterroso C, Santos-Ucha P, Prieto-Fernández Á (2013a) Phytoremediation of hexachlorocyclohexane (HCH)-contaminated soils using *Cytisus striatus* and bacterial inoculants in soils with distinct organic matter content. Environ Pollut 178:202–210. https://doi.org/10.1016/j.envpol.2013.03.027

Becerra-Castro C, Prieto-Fernández Á, Kidd PS, Weyens N, Rodríguez-Garrido B, Touceda-González M, Acea MJ, Vangronsveld J (2013b) Improving performance of *Cytisus striatus* on substrates contaminated with hexachlorocyclohexane (HCH) isomers using bacterial inoculants: developing a phytoremediation strategy. Plant and Soil 362:247–260. https://doi.org/10.1007/s11104-012-1276-6

Begum SR, Tamilselvi K (2016) Endophytes are plant helpers: an overview. Int J Curr Microbiol App Sci 5:424–436. https://doi.org/10.20546/ijcmas.2016.504.050
Bell TH, Joly S, Pitre FE, Yergeau E (2014) Increasing phytoremediation efficiency and reliability using novel omics approaches. Trends Biotechnol 32(5):271–280. https://doi.org/10.1016/j.tibtech.2014.02.008
Berg G (2009) Plant–microbe interactions promoting plant growth and health: perspectives for controlled use of microorganisms in agriculture. Appl Microbiol Biotechnol 84:11–18. https://doi.org/10.1007/s00253-009-2092-7
Berg G, Müller H, Zachow C, Opelt K, Scherwinski K, Tilcher R, Ulrich A, Hallmann J, Grosch R, Sessitsch A (2008) Endophytes: structural and functional diversity and biotechnological applications in control of plant pathogens. Ecol Genet 6:17–26. https://doi.org/10.17816/ecogen6217-26
Bhattacharyya PN, Jha DK (2012) Plant growth-promoting rhizobacteria (PGPR): emergence in agriculture. World J Microbiol Biotechnol 28:1327–1350. https://doi.org/10.1007/s11274-011-0979-9
Binet P, Portal JM, Leyval C (2000) Dissipation of 3–6-ring polycyclic aromatic hydrocarbons in the rhizosphere of ryegrass. Soil Biol Biochem 32:2011–2017. https://doi.org/10.1016/S0038-0717(00)00100-0
Bittsánszky A, Gullner G, Gyulai G, Komives T (2011) A case study: uptake and accumulation of persistent organic pollutants in *Cucurbitaceae* species. In: Schröder P, Collins CD (eds) Organic xenobiotics and plants: from mode of action to ecophysiology. Springer, Dordrecht, pp 77–85. https://doi.org/10.1007/978-90-481-9852-8_4
Bloemberg GV, Lugtenberg BJJ (2001) Molecular basis of plant growth promotion and biocontrol by rhizobacteria. Curr Opin Plant Biol 4:343–350. https://doi.org/10.1016/S1369-5266(00)00183-7
Bloemberg GV, Wijfjes AHM, Lamers GEM, Stuurman N, Lugtenberg BJJ (2000) Simultaneous imaging of *Pseudomonas fluorescens* WCS365 populations expressing three different autofluorescent proteins in the rhizosphere: new perspectives for studying microbial communities. Mol Plant Microbe Interact 13:1170–1176. https://doi.org/10.1094/mpmi.2000.13.11.1170
Brader G, Compant S, Mitter B, Trognitz F, Sessitsch A (2014) Metabolic potential of endophytic bacteria. Curr Opin Biotechnol 27:30–37. https://doi.org/10.1016/j.copbio.2013.09.012
Brader G, Compant S, Vescio K, Mitter B, Trognitz F, Ma L-J, Sessitsch A (2017) Ecology and genomic insights into plant-pathogenic and plant-nonpathogenic endophytes. Annu Rev Phytopathol 55:61–83. https://doi.org/10.1146/annurev-phyto-080516-035641
Burken JG (2004) Uptake and metabolism of organic compounds: green-liver model. In: McCutcheon SC, Schnoor JL (eds) Phytoremediation: transformation and control of contaminants. Wiley, Hoboken, pp 59–84. https://doi.org/10.1002/047127304X.ch2
Burken JG, Shanks JV, Thompson PL (2000) Phytoremediation and plant metabolism of explosives and nitroaromatic compounds. In: Spain JC, Hughes JB, Knackmuss H-J (eds) Biodegradation of nitroaromatic compounds and explosives, 1st edn. CRC Press LLC, Boca Raton, pp 239–276
Cardinale M (2014) Scanning a microhabitat: plant-microbe interactions revealed by confocal laser microscopy. Front Microbiol 5. https://doi.org/10.3389/fmicb.2014.00094
Cardinale M, Berg G (2015) Visualization of plant-microbe interactions. In: Lugtenberg B (ed) Principles of plant-microbe interactions: microbes for sustainable agriculture. Springer, Cham, pp 299–306. https://doi.org/10.1007/978-3-319-08575-3_31
Chaîneau CH, Morel JL, Oudot J (1997) Phytotoxicity and plant uptake of fuel oil hydrocarbons. J Environ Qual 26:1478–1483. https://doi.org/10.2134/jeq1997.00472425002600060005x
Chandra P, Singh E (2016) Applications and mechanisms of plant growth-stimulating rhizobacteria. In: Choudhary D, Varma A, Tuteja N (eds) Plant-microbe interaction: an approach to sustainable agriculture. Springer, Singapore, pp 37–62
Chanway CP (1996) Endophytes: they're not just fungi! Can J Bot 74:321–322
Chaturvedi H, Singh V, Gupta G (2016) Potential of bacterial endophytes as plant growth promoting factors. J Plant Pathol 7:7–9. https://doi.org/10.4172/2157-7471.1000376

Chen Y, Murrell JC (2010) When metagenomics meets stable-isotope probing: progress and perspectives. Trends Microbiol 18:157–163. https://doi.org/10.1016/j.tim.2010.02.002
Cherian S, Oliveira MM (2005) Transgenic plants in phytoremediation: recent advances and new possibilities. Environ Sci Technol 39:9377–9390. https://doi.org/10.1021/es051134l
Chikere CB, Okpokwasili GC, Chikere BO (2011) Monitoring of microbial hydrocarbon remediation in the soil. 3 Biotech 1:117–138. https://doi.org/10.1007/s13205-011-0014-8
Chiou CT, Sheng G, Manes M (2001) A partition-limited model for the plant uptake of organic contaminants from soil and water. Environ Sci Technol 35:1437–1444. https://doi.org/10.1021/es0017561
Coenye T, Vandamme P (2003) Diversity and significance of *Burkholderia* species occupying diverse ecological niches. Environ Microbiol 5:719–729. https://doi.org/10.1046/j.1462-2920.2003.00471.x
Collins CD (2008) A semi-quantitative approach to deriving a model structure for the uptake of organic chemicals by vegetation. Int J Phytoremediation 10:371–377. https://doi.org/10.1080/15226510802100440
Collins CD, Fryer M, Grosso A (2006) Plant uptake of non-ionic organic chemicals. Environ Sci Technol 40:45–52. https://doi.org/10.1021/es0508166
Compant S, Duffy B, Nowak J, Clément C, Barka EA (2005) Use of plant growth-promoting bacteria for biocontrol of plant diseases: principles, mechanisms of action, and future prospects. Appl Environ Microbiol 71:4951–4959. https://doi.org/10.1128/AEM.71.9.4951-4959.2005
Compant S, Kaplan H, Sessitsch A, Nowak J, Ait Barka E, Clément C (2008) Endophytic colonization of *Vitis vinifera* L. by *Burkholderia phytofirmans* strain PsJN: from the rhizosphere to inflorescence tissues. FEMS Microbiol Ecol 63:84–93. https://doi.org/10.1111/j.1574-6941.2007.00410.x
Compant S, Clément C, Sessitsch A (2010) Plant growth-promoting bacteria in the rhizo- and endosphere of plants: their role, colonization, mechanisms involved and prospects for utilization. Soil Biol Biochem 42:669–678. https://doi.org/10.1016/j.soilbio.2009.11.024
Compant S, Sessitsch A, Mathieu F (2012) The 125th anniversary of the first postulation of the soil origin of endophytic bacteria-a tribute to M.L.V. Galippe. Plant and Soil 356:299–301. https://doi.org/10.1007/s11104-012-1204-9
Compant S, Saikkonen K, Mitter B, Campisano A, Mercado-Blanco J (2016) Editorial special issue: soil, plants and endophytes. Plant and Soil 405:1–11. https://doi.org/10.1007/s11104-016-2927-9
Compant S, Samad A, Faist H et al A review on the plant microbiome: ecology, functions, and emerging trends in microbial application. J Adv Res. https://doi.org/10.1016/j.jare.2019.03.004
De Bary A (1879) Die erscheinung der symbiose. Verlag Trubner, Strassburg
de Mello-Farias PC, ALS C, Lencina CL (2011) Transgenic plants for enhanced phytoremediation–physiological studies. In: Alvarez M (ed) Genetic transformation. InTech, Rijeka, pp 305–328. https://doi.org/10.5772/24355
Dhankher OP, Pilon-Smits EAH, Meagher RB, Doty S (2012) Biotechnological approaches for phytoremediation. In: Altman A, Hasegawa PM (eds) Plant biotechnology and agriculture. Academic Press, San Diego, pp 309–328. https://doi.org/10.1016/B978-0-12-381466-1.00020-1
Dietz AC, Schnoor JL (2001) Advances in phytoremediation. Environ Health Perspect 109:163–168. https://doi.org/10.2307/3434854
Dong R, Gu L, Guo C, Xun F, Liu J (2014) Effect of PGPR *Serratia marcescens* BC-3 and AMF *Glomus intraradices* on phytoremediation of petroleum contaminated soil. Ecotoxicology 23:674–680. https://doi.org/10.1007/s10646-014-1200-3
Doty SL (2008) Enhancing phytoremediation through the use of transgenics and endophytes. New Phytol 179:318–333. https://doi.org/10.1111/j.1469-8137.2008.02446.x
Doty SL (2017a) Endophytic N-fixation: controversy and a path forward. In: Doty SL (ed) Functional importance of the plant microbiome: implications for agriculture, forestry and bioenergy. Springer, Cham, pp 7–20

Doty SL (2017b) Functional importance of the plant endophytic microbiome: implications for agriculture, forestry, and bioenergy. In: Doty SL (ed) Functional importance of the plant microbiome: implications for agriculture, forestry and bioenergy. Springer, Cham, pp 1–5

Doty SL, Shang TQ, Wilson AM, Tangen J, Westergreen AD, Newman LA, Strand SE, Gordon MP (2000) Enhanced metabolism of halogenated hydrocarbons in transgenic plants containing mammalian cytochrome P450 2E1. Proc Natl Acad Sci U S A 97:6287–6291. https://doi.org/10.1073/pnas.97.12.6287

Doty SL, James CA, Moore AL, Vajzovic A, Singleton GL, Ma C, Khan Z, Xin G, Kang JW, Park JY, Meilan R, Strauss SH, Wilkerson J, Farin F, Strand SE (2007) Enhanced phytoremediation of volatile environmental pollutants with transgenic trees. Proc Natl Acad Sci U S A 104(43):16816–16821. https://doi.org/10.1073/pnas.0703276104

Doty SL, Khan Z, Kang JW, Knoth J, Roman D, Morrison H, Baum A (2012) Endophytes for enhanced plant growth, pollutant degradation and biofuels. In: In vitro cellular & developmental biology-animal, vol 48, Suppl. 1. Springer, New York, p 21. https://doi.org/10.1007/s11626-012-9506-z

Doty SL, Freeman JL, Cohu CM, Burken JG, Firrincieli A, Simon A, Khan Z, Isebrands JG, Lukas J, Blaylock MJ (2017) Enhanced degradation of TCE on a superfund site using endophyte-assisted poplar tree phytoremediation. Environ Sci Technol 51:10050–10058. https://doi.org/10.1021/acs.est.7b01504

Eapen S, Singh S, D'Souza SF (2007) Advances in development of transgenic plants for remediation of xenobiotic pollutants. Biotechnol Adv 25:442–451. https://doi.org/10.1016/j.biotechadv.2007.05.001

Eevers N, Van Hamme JD, Bottos EM, Weyens N, Vangronsveld J (2015a) Draft genome sequence of *Enterobacter aerogenes*, a DDE-degrading and plant growth-promoting strain isolated from *Cucurbita pepo*. Genome Announc 3. https://doi.org/10.1128/genomeA.00317-15

Eevers N, Van Hamme JD, Bottos EM, Weyens N, Vangronsveld J (2015b) *Sphingomonas taxi*, isolated from *Cucurbita pepo*, proves to be a DDE-degrading and plant growth-promoting strain. Genome Announc 3. https://doi.org/10.1128/genomeA.00489-15

Eevers N, Van Hamme JD, Bottos EM, Weyens N, Vangronsveld J (2015c) Draft genome sequence of *Methylobacterium radiotolerans*, a DDE-degrading and plant growth-promoting strain isolated from *Cucurbita pepo*. Genome Announc 3. https://doi.org/10.1128/genomeA.00488-15

Eevers N, Gielen M, Sánchez-López A, Jaspers S, White JC, Vangronsveld J, Weyens N (2015d) Optimization of isolation and cultivation of bacterial endophytes through addition of plant extract to nutrient media. J Microbial Biotechnol 8:707–715. https://doi.org/10.1111/1751-7915.12291

Eevers N, Hawthorne JR, White JC, Vangronsveld J, Weyens N (2016) Exposure of *Cucurbita pepo* to DDE-contamination alters the endophytic community: a cultivation dependent vs a cultivation independent approach. Environ Pollut 209:147–154. https://doi.org/10.1016/j.envpol.2015.11.038

Eevers N, White JC, Vangronsveld J, Weyens N (2017) Bio- and phytoremediation of pesticide-contaminated environments: a review. In: Cuypers A, Vangronsveld J (eds) Advances in botanical research, vol 83. Academic Press, London, pp 277–318. https://doi.org/10.1016/bs.abr.2017.01.001

Eevers N, Hawthorne JR, White JC, Vangronsveld J, Weyens N (2018) Endophyte-enhanced phytoremediation of DDE-contaminated using *Cucurbita pepo*: A field trial, Int J Phytoremediation 20:4, 301–310. https://doi.org/10.1080/15226514.2017.1377150

Eskandary S, Tahmourespour A, Hoodaji M, Abdollahi A (2017) The synergistic use of plant and isolated bacteria to clean up polycyclic aromatic hydrocarbons from contaminated soil. J Environ Health Sci Eng 15(12). https://doi.org/10.1186/s40201-017-0274-2

Etesami H, Alikhani HA, Hosseini HM (2015) Indole-3-acetic acid (IAA) production trait, a useful screening to select endophytic and rhizosphere competent bacteria for rice growth promoting agents. MethodsX 2:72–78. https://doi.org/10.1016/j.mex.2015.02.008

Feng F, Ge J, Li Y, Cheng J, Zhong J, Yu X (2017a) Isolation, colonization, and chlorpyrifos degradation mediation of the endophytic bacterium *Sphingomonas* strain HJY in Chinese chives (*Allium tuberosum*). J Agric Food Chem 65:1131–1138. https://doi.org/10.1021/acs.jafc.6b0528

Feng F, Ge J, Li Y, He S, Zhong J, Liu X, Yu X (2017b) Enhanced degradation of chlorpyrifos in rice (*Oryza sativa* L.) by five strains of endophytic bacteria and their plant growth promotional ability. Chemosphere 184:505–513. https://doi.org/10.1016/j.chemosphere.2017.05.178

Feng F, Li Y, Ge J, Chen J, Jiang W, He S, Liu X, Yu X (2017c) Degradation of chlorpyrifos by an endophytic bacterium of the *Sphingomonas* genus (strain HJY) isolated from Chinese chives (*Allium tuberosum*). J Environ Sci Health B 2017:1–9. https://doi.org/10.1080/03601234.2017.1356675

Feng N-X, Yu J, Zhao H-M, Cheng Y-T, Mo C-H, Cai Q-Y, Li Y-W, Li H, Wong M-H (2017d) Efficient phytoremediation of organic contaminants in soils using plant–endophyte partnerships. Sci Total Environ 583:352–368. https://doi.org/10.1016/j.scitotenv.2017.01.075

Fester T, Giebler J, Wick LY, Schlosser D, Kästner M (2014) Plant-microbe interactions as drivers of ecosystem functions relevant for the biodegradation of organic contaminants. Curr Opin Biotechnol 27:168–175. https://doi.org/10.1016/j.copbio.2014.01.017

Fismes J, Perrin-Ganier C, Empereur-Bissonnet P, Morel JL (2002) Soil-to-root transfer and translocation of polycyclic aromatic hydrocarbons by vegetables grown on industrial contaminated soils. J Environ Qual 31:1649–1656. https://doi.org/10.2134/jeq2002.1649

French CE, Rosser SJ, Davies GJ, Nicklin S, Bruce NC (1999) Biodegradation of explosives by transgenic plants expressing pentaerythritol tetranitrate reductase. Nat Biotechnol 17:491. https://doi.org/10.1038/8673

Gagne S, Richard C, Antoun H (1989) Pourvoir pathogne des bactries endoracinaires de la luzerne. Can J Plant Pathol 11:22–27

Gaiero JR, McCall CA, Thompson KA, Day NJ, Best AS, Dunfield KE (2013) Inside the root microbiome: bacterial root endophytes and plant growth promotion. Am J Bot 100:1738–1750. https://doi.org/10.3732/ajb.1200572

Galippe V (1887a) Note sur la présence de micro-organismes dans les tissus végétaux. Comptes Rendus Hebdomadaires de la Société de Biologie, Paris, pp 410–416

Galippe V (1887b) Note sur la présence de micro-organismes dans les tissus végétaux (deuxième note) CR. Seances Soc Biol Fil 39:557–560

Gamalero E, Glick BR (2011) Mechanisms used by plant growth-promoting bacteria. In: Maheshwari DK (ed) Bacteria in agrobiology: plant nutrient management. Springer, Heidelberg, pp 17–46. https://doi.org/10.1007/978-3-642-21061-7_2

Gao Y, Zhu L (2004) Plant uptake, accumulation and translocation of phenanthrene and pyrene in soils. Chemosphere 55(9):1169–1178. https://doi.org/10.1016/j.chemosphere.2004.01.037

Gao Y, Cheng Z, Ling W, Huang J (2010) Arbuscular mycorrhizal fungal hyphae contribute to the uptake of polycyclic aromatic hydrocarbons by plant roots. Bioresour Technol 101:6895–6901. https://doi.org/10.1016/j.biortech.2010.03.122

Gelvin SB (2003) *Agrobacterium*-mediated plant transformation: the biology behind the "gene-jockeying" tool. Microbiol Mol Biol Rev 67:16–37. https://doi.org/10.1128/MMBR.67.1.16-37.2003

Gerhardt KE, Huang X-D, Glick BR, Greenberg BM (2009) Phytoremediation and rhizoremediation of organic soil contaminants: potential and challenges. Plant Sci 176:20–30. https://doi.org/10.1016/j.plantsci.2008.09.014

Gerhardt KE, Gerwing PD, Huang X, Greenberg BM (2015) Microbe-assisted phytoremediation of petroleum impacted soil. In: Fingas M (ed) Handbook of oil spill science and technology, 1st edn. Wiley, pp 407–427. https://doi.org/10.1002/9781118989982.ch16

Gerhardt KE, Gerwing PD, Greenberg BM (2017) Opinion: taking phytoremediation from proven technology to accepted practice. Plant Sci 256:170–185. https://doi.org/10.1016/j.plantsci.2016.11.016

Germaine KJ, Liu X, Cabellos GG, Hogan JP, Ryan D, Dowling DN (2006) Bacterial endophyte-enhanced phytoremediation of the organochlorine herbicide 2,4-dichlorophenoxyacetic acid. FEMS Microbiol Ecol 57:302–310. https://doi.org/10.1111/j.1574-6941.2006.00121.x
Germaine KJ, Keogh E, Ryan D, Dowling DN (2009) Bacterial endophyte-mediated naphthalene phytoprotection and phytoremediation. FEMS Microbiol Lett 296:226–234. https://doi.org/10.1111/j.1574-6968.2009.01637.x
Gkorezis P (2014) The role of plant-associated bacteria in the remediation of soils and groundwater contaminated with petroleum hydrocarbons. (Doctoral dissertation). Faculty of Sciences, Hasselt University (Universiteit Hasselt), Diepenbeek, Belgium. http://hdl.handle.net/1942/20493
Gkorezis P, Daghio M, Franzetti A, Van Hamme JD, Sillen W, Vangronsveld J (2016) The interaction between plants and bacteria in the remediation of petroleum hydrocarbons: an environmental perspective. Front Microbiol 7:1–27. https://doi.org/10.3389/fmicb.2016.01836
Glick BR (2003) Phytoremediation: synergistic use of plants and bacteria to clean up the environment. Biotechnol Adv 21:383–393. https://doi.org/10.1016/S0734-9750(03)00055-7
Glick BR (2010) Using soil bacteria to facilitate phytoremediation. Biotechnol Adv 28:367–374. https://doi.org/10.1016/j.biotechadv.2010.02.001
Glick BR (2012) Plant growth-promoting bacteria: mechanisms and applications. Scientifica 2012:15. https://doi.org/10.1016/j.micres.2013.09.009
Glick BR (2014) Bacteria with ACC deaminase can promote plant growth and help to feed the world. Microbiol Res 169:30–39. https://doi.org/10.1016/j.micres.2013.09.009
Glick BR (2015) Resource acquisition. In: Glick BR (ed) Beneficial plant-bacterial interactions. Springer, Cham, pp 29–63. https://doi.org/10.1007/978-3-319-13921-0_2
Glick BR, Stearns JC (2011) Making phytoremediation work better: maximizing a plant's growth potential in the midst of adversity. Int J Phytoremediation 13:4–16. https://doi.org/10.1080/15226514.2011.568533
Goodwin S, Mcpherson JD, McCombie WR (2016) Coming of age: ten years of next-generation sequencing technologies. Nat Rev Genet 17(6):333–351. https://doi.org/10.1038/nrg.2016.49
Gray EJ, Smith DL (2005) Intracellular and extracellular PGPR: commonalities and distinctions in the plant–bacterium signaling processes. Soil Biol Biochem 37:395–412. https://doi.org/10.1016/j.soilbio.2004.08.030
Gupta G, Parihar SS, Ahirwar NK, Snehi SK, Singh V (2015) Plant growth promoting rhizobacteria (PGPR): current and future prospects for development of sustainable agriculture. J Microb Biochem Technol 7:096–102. https://doi.org/10.4172/1948-5948.1000188
Hallmann J (2001) Plant interactions with endophytic bacteria. In: Jeger MJ, Spence NJ (eds) Biotic interactions in plant-pathogen associations. CABI, New York, pp 87–119
Hallmann J, Berg G (2006) Spectrum and population dynamics of bacterial root endophytes. In: Schulz BJE, Boyle CJC, Sieber TN (eds) Microbial root endophytes (soil biology series). Springer, Heidelberg, pp 15–31. https://doi.org/10.1007/3-540-33526-9_2
Hallmann J, Quadt-Hallmann A, Mahaffee WF, Kloepper JW (1997) Bacterial endophytes in agricultural crops. Can Microbiol 43:895–914. https://doi.org/10.1139/m97-131
Hardoim PR, van Elsas JD (2013) Properties of bacterial endophytes leading to maximized host fitness. In: de Bruijn FJ (ed) Molecular microbial ecology of the rhizosphere, vol 1 & 2. Wiley, Hoboken, pp 405–411. https://doi.org/10.1002/9781118297674.ch37
Hardoim PR, van Overbeek LS, Elsas JD (2008) Properties of bacterial endophytes and their proposed role in plant growth. Trends Microbiol 16:463–471. https://doi.org/10.1016/j.tim.2008.07.008
Hardoim PR, van Overbeek LS, Berg G, Pirttilä AM, Compant S, Campisano A, Döring M, Sessitsch A (2015) The hidden world within plants: ecological and evolutionary considerations for defining functioning of microbial endophytes. Microbiol Mol Biol Rev 79:293–320. https://doi.org/10.1128/MMBR.00050-14
Harish S, Kavino M, Kumar N, Balasubramanian P, Samiyappan R (2009) Induction of defense-related proteins by mixtures of plant growth promoting endophytic bacteria against Banana bunchy top virus. Biol Control 51:16–25. https://doi.org/10.1016/j.biocontrol.2009.06.002

Harms H, Bokern M, Kolb M, Bock C (2004) Transformation of organic contaminants by different plant systems. In: McCutcheon SC, Schnoor JL (eds) Phytoremediation: transformation and control of contaminants. Wiley, Hoboken, pp 285–316. https://doi.org/10.1002/047127304X.ch9

Hartmann A, Schmid M, van Tuinen D, Berg G (2009) Plant-driven selection of microbes. Plant and Soil 321:235. https://doi.org/10.1007/s11104-008-9814-y

Hellström A (2004) Uptake of organic pollutants in plants. Dissertation, Swedish University of Agricultural Sciences, Uppsala

Hentschel U, Steinert M, Hacker J (2000) Common molecular mechanisms of symbiosis and pathogenesis. Trends Microbiol 8:226–231. https://doi.org/10.1016/S0966-842X(00)01758-3

Ho Y-N, Shih C-H, Hsiao S-C, Huang C-C (2009) A novel endophytic bacterium, *Achromobacter xylosoxidans*, helps plants against pollutant stress and improves phytoremediation. J Biosci Bioeng 108:S94. https://doi.org/10.1016/j.jbiosc.2009.08.276

Ho Y-N, Mathew DC, Hsiao S-C, Shih C-H, Chien M-F, Chiang H-M, Huang C-C (2012) Selection and application of endophytic bacterium *Achromobacter xylosoxidans* strain F3B for improving phytoremediation of phenolic pollutants. J Hazard Mater 219-220:43–49. https://doi.org/10.1016/j.jhazmat.2012.03.035

Ho Y-N, Hsieh J-L, Huang C-C (2013) Construction of a plant–microbe phytoremediation system: combination of vetiver grass with a functional endophytic bacterium, *Achromobacter xylosoxidans* F3B, for aromatic pollutants removal. Bioresour Technol 145:43–47. https://doi.org/10.1016/j.biortech.2013.02.051

Hollis JP (1949) Bacteria in healthy potato tissue. ETD collection for University of Nebraska – Lincoln. AAIDP13795. http://digitalcommons.unl.edu/dissertations/AAIDP13795

Hollis JP (1951) Bacteria in healthy potato tissue. Phytopathology 41:350–366

Howe D, Costanzo M, Fey P, Gojobori T, Hannick L, Hide W, Hill DP, Kania R, Schaeffer M, St Pierre S, Twigger S, White O, Rhee SY (2008) Big data: the future of biocuration. Nature 455:47–50. https://doi.org/10.1038/455047a

Huss-Danell K (1997) Actinorhizal symbioses and their N_2 fixation, vol 136, pp 375–405. https://doi.org/10.1046/j.1469-8137.1997.00755.x

Imadi S, Shazadi K, Gul A, Hakeem K (2015) Sustainable crop production system. In: Hakeem KR, Akhtar MS, Abdullah SNA (eds) Plant, soil and microbes, implications in crop science, vol 1. Springer, Cham, pp 103–116. https://doi.org/10.1007/978-3-319-27455-3_6

Jabeen H, Iqbal S, Ahmad F, Afzal M, Firdous S (2016) Enhanced remediation of chlorpyrifos by ryegrass (*Lolium multiflorum*) and a chlorpyrifos degrading bacterial endophyte *Mezorhizobium* sp. HN3. Int J Phytoremediation 18:126–133. https://doi.org/10.1080/15226514.2015.1073666

Jambon I, Thijs S, Weyens N, Vangronsveld J (2018) Harnessing plant-bacteria-fungi interactions to improve plant growth and degradation of organic pollutants. J Plant Interact 13:119–130. https://doi.org/10.1080/17429145.2018.1441450

James EK (2000) Nitrogen fixation in endophytic and associative symbiosis. Field Crop Res 65:197–209. https://doi.org/10.1016/S0378-4290(99)00087-8

Jørgensen TS, Kiil AS, Hansen MA, Sørensen SJ, Hansen LH (2015) Current strategies for mobilome research. Front Microbiol 5. https://doi.org/10.3389/fmicb.2014.00750

Kacálková L, Tlustoš P (2011) The uptake of persistent organic pollutants by plants. Open Life Sci 6(2):223–235. https://doi.org/10.2478/s11535-010-0116-z

Kandel S, Joubert P, Doty S (2017) Bacterial endophyte colonization and distribution within plants. Microorganisms 5:77. https://doi.org/10.3390/microorganisms5040077

Kaul S, Sharma T, Dhar MK (2016) "Omics" tools for better understanding the plant–endophyte interactions. Front Plant Sci 7(955). https://doi.org/10.3389/fpls.2016.00955

Kawahigashi H (2009) Transgenic plants for phytoremediation of herbicides. Curr Opin Biotechnol 20:225–230. https://doi.org/10.1016/j.copbio.2009.01.010

Khan Z, Doty S (2011) Endophyte-assisted phytoremediation. Plant Biol 12:97–105

Khan S, Afzal M, Iqbal S, Khan QM (2013a) Plant–bacteria partnerships for the remediation of hydrocarbon contaminated soils. Chemosphere 90:1317–1332. https://doi.org/10.1016/j.chemosphere.2012.09.045

Khan S, Afzal M, Iqbal S, Mirza MS, Khan QM (2013b) Inoculum pretreatment affects bacterial survival, activity and catabolic gene expression during phytoremediation of diesel contaminated soil. Chemosphere 91:663–668. https://doi.org/10.1016/j.chemosphere.2013.01.025

Kloepper JW, Ryu C-M (2006) Bacterial endophytes as elicitors of induced systemic resistance. In: Schulz BJE, Boyle CJC, Sieber TN (eds) Microbial root endophytes. Springer, Heidelberg, pp 33–52. https://doi.org/10.1007/3-540-33526-9_3

Kloepper JW, Schippers B, Bakker PAHM (1992) Proposed elimination of the term "endorhizosphere". Phytopathology 82:726–727

Krishnaraj PU, Pasha MA (2017) Metagenome of rhizosphere and endophytic ecosystem. In: Singh RP, Kothari R, Koringa PG, Singh SP (eds) Understanding host-microbiome interactions – an omics approach. Springer, Singapore, pp 125–156. https://doi.org/10.1007/978-981-10-5050-3_9

Kuiper I, Bloemberg GV, Lugtenberg BJ (2001) Selection of a plant-bacterium pair as a novel tool for rhizostimulation of polycyclic aromatic hydrocarbon-degrading bacteria. Mol Plant Microbe Interact 14:1197–1205. https://doi.org/10.1094/mpmi.2001.14.10.1197

Kuiper I, Kravchenko LV, Bloemberg GV, Lugtenberg BJ (2002) Pseudomonas putida strain PCL1444, selected for efficient root colonization and naphthalene degradation, effectively utilizes root exudate components. Mol Plant Microbe Interact 15:734–741. https://doi.org/10.1094/mpmi.2002.15.7.734

Kukla M, Płociniczak T, Piotrowska-Seget Z (2014) Diversity of endophytic bacteria in *Lolium perenne* and their potential to degrade petroleum hydrocarbons and promote plant growth. Chemosphere 117:40–46. https://doi.org/10.1016/j.chemosphere.2014.05.055

Kvesidatza E, Sadunishvili T, Kvesitadze G (2009) Mechanisms of organic contaminants uptake and degradation in plants. World Acad Sci Eng Technol 55(6):458–468

Kvesitadze G, Gordeziani M, Khatisashvili G, Sadunishvili T, Ramsden J (2001) Some aspects of the enzymatic basis of phytoremediation. J Biol Phys Chem 1:49–57. https://doi.org/10.4024/16KV01R.01.02

Lacava PT, Azevedo JL (2013) Endophytic bacteria: a biotechnological potential in agrobiology system. In: Maheshwari D, Saraf M, Aeron A (eds) Bacteria in agrobiology: crop productivity. Springer, Heidelberg, pp 1–44. https://doi.org/10.1007/978-3-642-37241-4_1

Lagendijk EL, Lamers GEM, Bloemberg GV, Validov S, De Weert S (2010) Genetic tools for tagging Gram-negative bacteria with mCherry for visualization *in vitro* and in natural habitats, biofilm and pathogenicity studies. FEMS Microbiol Lett 305:81–90. https://doi.org/10.1111/j.1574-6968.2010.01916.x

Li QQ, Loganath A, Chong YS, Tan J, Obbard JP (2006) Persistent organic pollutants and adverse health effects in humans. J Toxicol Environ Health A 69:1987–2005. https://doi.org/10.1080/15287390600751447

Li K, Pidatala VR, Shaik R, Datta R, Ramakrishna W (2014) Integrated metabolomic and proteomic approaches dissect the effect of metal-resistant bacteria on maize biomass and copper uptake. Environ Sci Technol 48:1184. https://doi.org/10.1021/es4047395

Limmer M, Burken J (2016) Phytovolatilization of organic contaminants. Environ Sci Technol 50 (13):6632–6643. https://doi.org/10.1021/acs.est.5b04113

Liu H, Carvalhais LC, Crawford M, Singh E, Dennis PG, Pieterse CMJ, Schenk PM (2017) Inner plant values: diversity, colonization and benefits from endophytic bacteria. Front Microbiol 8 (2552). https://doi.org/10.3389/fmicb.2017.02552

Liu J, Zhang Z, Sheng Y, Gao Y, Zhao Z (2018) Phenanthrene-degrading bacteria on root surfaces: a natural defense that protects plants from phenanthrene contamination. Plant Soil. 1–16. https://doi.org/10.1007/s11104-018-3575-z

Lodewyckx C, Vangronsveld J, Porteous F, Moore ERB, Taghavi S, Mezgeay M, der Lelie DV (2002) Endophytic bacteria and their potential applications. Crit Rev Plant Sci 21:583–606. https://doi.org/10.1080/0735-260291044377

Loman NJ, Constantinidou C, Chan JZM, Halachev M, Sergeant M, Penn CW, Robinson ER, Pallen MJ (2012a) High-throughput bacterial genome sequencing: an embarrassment of choice, a world of opportunity. Nat Rev Microbiol 10:599–606. https://doi.org/10.1038/nrmicro2850

Loman NJ, Misra RV, Dallman TJ, Constantinidou C, Gharbia SE, Wain J, Pallen MJ (2012b) Performance comparison of benchtop high-throughput sequencing platforms. Nat Biotechnol 30:434–439. https://doi.org/10.1038/nbt.2198

Lugtenberg BJ, Chin-A-Woeng TF, Bloemberg GV (2002) Microbe–plant interactions: principles and mechanisms. Antonie Van Leeuwenhoek 81:373–383. https://doi.org/10.1023/a:1020596903142

Lumactud R, Fulthorpe RR (2018) Endophytic bacterial community structure and function of herbaceous plants from petroleum hydrocarbon contaminated and non-contaminated sites. Front Microbiol 9(1926). https://doi.org/10.3389/fmicb.2018.01926

Lumactud R, Shen SY, Lau M, Fulthorpe R (2016) Bacterial endophytes isolated from plants in natural oil seep soils with chronic hydrocarbon contamination. Front Microbiol 7:1–10. https://doi.org/10.3389/fmicb.2016.00755

Ma X, Burken JG (2003) TCE diffusion to the atmosphere in phytoremediation applications. Environ Sci Technol 37:2534–2539. https://doi.org/10.1021/es026055d

Macek T, Macková M, Káš J (2000) Exploitation of plants for the removal of organics in environmental remediation. Biotechnol Adv 18:23–34. https://doi.org/10.1016/S0734-9750(99)00034-8

Macek T, Kotrba P, Svatos A, Novakova M, Demnerova K, Mackova M (2008) Novel roles for genetically modified plants in environmental protection. Trends Biotechnol 26:146–152. https://doi.org/10.1016/j.tibtech.2007.11.009

Martin FM, Uroz S, Barker DG (2017) Ancestral alliances: plant mutualistic symbioses with fungi and bacteria. Science 356(6340). https://doi.org/10.1126/science.aad4501

Mastretta C, Barac T, Vangronsveld J, Newman L, Taghavi S, Lelie DVD (2006) Endophytic bacteria and their potential application to improve the phytoremediation of contaminated environments. Biotechnol Genet Eng Rev 23:175–188. https://doi.org/10.1080/02648725.2006.10648084

McCully ME (2001) Niches for bacterial endophytes in crop plants: a plant biologist's view. Aust J Plant Physiol 28(9):983–990. https://doi.org/10.1071/PP01101

McGuinness MC, Mazurkiewicz V, Brennan E, Dowling DN (2007) Dechlorination of pesticides by a specific bacterial glutathione-*S*-transferase, $BphK^{LB400}$: potential for bioremediation. Eng Life Sci 7:611–615. https://doi.org/10.1002/elsc.200720218

McKone TE, Maddalena RL (2007) Plant uptake of organic pollutants from soil: bioconcentration estimates based on models and experiments. Environ Toxicol Chem 26:2494–2504. https://doi.org/10.1897/06-269.1

Meena KK, Sorty AM, Bitla UM, Choudhary K, Gupta P, Pareek A, Singh DP, Prabha R, Sahu PK, Gupta VK, Singh HB, Krishanani KK, Minhas PS (2017) Abiotic stress responses and microbe-mediated mitigation in plants: the omics strategies. Front Plant Sci 8. https://doi.org/10.3389/fpls.2017.00172

Megharaj M, Ramakrishnan B, Venkateswarlu K, Sethunathan N, Naidu R (2011) Bioremediation approaches for organic pollutants: a critical perspective. Environ Int 37:1362–1375. https://doi.org/10.1016/j.envint.2011.06.003

Mercado-Blanco J, Lugtenberg B (2014) Biotechnological applications of bacterial endophytes. Curr Biotechnol 3:60–75. https://doi.org/10.2174/22115501113026660038

Mitter B, Petric A, Shin M, Chain PSG, Hauberg-Lotte L, Reinhold-Hurek B, Nowak J, Sessitsch A (2013a) Comparative genome analysis of *Burkholderia phytofirmans* PsJN reveals a wide spectrum of endophytic lifestyles based on interaction strategies with host plants. Front Plant Sci 4. https://doi.org/10.3389/fpls.2013.00120

Mitter B, Petric A, Chain PSG, Trognitz F, Nowak J, Compant S, Sessitsch A (2013b) Genome analysis, ecology, and plant growth promotion of the endophyte *Burkholderia phytofirmans* strain PsJN. In: de Bruijn FJ (ed) Molecular microbial ecology of the rhizosphere, vol 1 & 2. Wiley, Hoboken, pp 865–874. https://doi.org/10.1002/9781118297674.ch81

Mitter B, Brader G, Afzal M, Compant S, Naveed M, Trognitz F, Sessitsch A (2013c) Advances in elucidating beneficial interactions between plants, soil, and bacteria. In: Sparks DL (ed) Advances in agronomy, vol 121. Academic Press, Burlington, pp 381–445

Montpetit É, Lachapelle E (2017) New environmental technology uptake and bias toward the status quo: the case of phytoremediation. Environ Technol Innov 7:102–109. https://doi.org/10.1016/j.eti.2016.12.008

Morsy M (2015) Microbial symbionts: a potential bio-boom. J Investig Genet 2:00015. https://doi.org/10.15406/jig.2015.02.00015

Mulas D, Díaz-Alcántara CA, Mulas R, Marcano I, Barquero M, Serrano P, González-Andrés F (2013) Inoculants based in autochthonous microorganisms, a strategy to optimize agronomic performance of biofertilizers. In: Gonzalez BR, Gonzalez-Lopez J (eds) Beneficial plant-microbial interactions: ecology and applications, 1st edn. CRC Press, Boca Raton, pp 309–337. https://doi.org/10.1201/b15251-14

Müller CA, Obermeier MM, Berg G (2016) Bioprospecting plant-associated microbiomes. J Biotechnol 235:171–180. https://doi.org/10.1016/j.jbiotec.2016.03.033

Nadal M, Domingo JL (2013) Sources of human exposure. In: Carpenter J (ed) Effects of persistent and bioactive organic pollutants on human health. Wiley, New York, pp 8–25. https://doi.org/10.1002/9781118679654.ch2

Nekrutenko A, Taylor J (2012) Next-generation sequencing data interpretation: enhancing reproducibility and accessibility. Nat Rev Genet 13(9):667–672. https://doi.org/10.1038/nrg3305

Newman LA, Reynolds CM (2005) Bacteria and phytoremediation: new uses for endophytic bacteria in plants. Trends Biotechnol 23:6–8. https://doi.org/10.1016/j.tibtech.2004.11.010

Newton AC, Fitt BDL, Atkins SD, Walters DR, Daniell TJ (2010) Pathogenesis, parasitism and mutualism in the trophic space of microbe-plant interactions. Trends Microbiol 18:365–373. https://doi.org/10.1016/j.tim.2010.06.002

Olanrewaju OS, Glick BR, Babalola OO (2017) Mechanisms of action of plant growth promoting bacteria. World J Microbiol Biotechnol 33(197). https://doi.org/10.1007/s11274-017-2364-9

Oldroyd GE, Murray JD, Poole PS, Downie JA (2011) The rules of engagement in the legume-rhizobial symbiosis. Annu Rev Genet 45:119–144. https://doi.org/10.1146/annurev-genet-110410-132549

Pandotra P, Raina M, Salgotra RK, Ali S, Mir ZA, Bhat JA, Tyagi A, Upadhahy D (2018) Plant-bacterial partnership: a major pollutants remediation approach. In: Oves M, Khan MZ, Ismail MII (eds) Modern age environmental problems and their remediation. Springer, Cham, pp 169–200. https://doi.org/10.1007/978-3-319-64501-8_10

Pawlik M, Cania B, Thijs S, Vangronsveld J, Piotrowska-Seget Z (2017) Hydrocarbon degradation potential and plant growth-promoting activity of culturable endophytic bacteria of *Lotus corniculatus* and *Oenothera biennis* from a long-term polluted site. Environ Sci Pollut Res Int 24:19640–19652. https://doi.org/10.1007/s11356-017-9496-1

Pereira L (2014) Persistent organic chemicals of emerging environmental concern. In: Malik A, Grohmann E, Akhtar R (eds) Environmental deterioration and human health: natural and anthropogenic determinants. Springer, Dordrecht, pp 163–213. https://doi.org/10.1007/978-94-007-7890-0_8

Perotti R (1926) On the limits of biological enquiry in soil science. Proc Int Soc Soil Sci 2:146–161

Phillips LA, Germida JJ, Farrell RE, Greer CW (2008) Hydrocarbon degradation potential and activity of endophytic bacteria associated with prairie plants. Soil Biol Biochem 40:3054–3064. https://doi.org/10.1016/j.soilbio.2008.09.006

Pilon-Smits E (2005) Phytoremediation. Annu Rev Plant Biol 56:15–39. https://doi.org/10.1146/annurev.arplant.56.032604.144214

Pilon-Smits EA, Freeman JL (2006) Environmental cleanup using plants: biotechnological advances and ecological considerations. Front Ecol Environ 4:203–210. https://doi.org/10.1890/1540-9295(2006)004[0203:ECUPBA]2.0.CO;2

Pińiski A, Hupert-Kocurek K (2016) Genetic basis of endophytic bacteria-plant interactions. Postep Mikrobiol 55(4):404–412

Pirttilä AM, Joensuu P, Pospiech H, Jalonen J, Hohtola A (2004) Bud endophytes of Scots pine produce adenine derivatives and other compounds that affect morphology and mitigate browning of callus cultures. Physiol Plant 121:305–312. https://doi.org/10.1111/j.0031-9317.2004.00330.x

Puri A, Padda KP, Chanway CP (2017) Nitrogen-fixation by endophytic bacteria in agricultural crops: recent advances, nitrogen in agriculture – updates, Amanullah and Shah Fahad, IntechOpen, https://doi.org/10.5772/intechopen.71988. https://www.intechopen.com/books/nitrogen-in-agriculture-updates/nitrogen-fixation-by-endophytic-bacteria-in-agricultural-crops-recent-advances

Rashid S, Charles TC, Glick BR (2012) Isolation and characterization of new plant growth-promoting bacterial endophytes. Appl Soil Ecol 61:217–224. https://doi.org/10.1016/j.apsoil.2011.09.011

Reichenauer TG, Germida JJ (2008) Phytoremediation of organic contaminants in soil and groundwater. ChemSusChem 1:708–717. https://doi.org/10.1002/cssc.200800125

Reinhold-Hurek B, Hurek T (2011) Living inside plants: bacterial endophytes. Curr Opin Plant Biol 14:435–443. https://doi.org/10.1016/j.pbi.2011.04.004

Reinhold-Hurek B, Bünger W, Burbano CS, Sabale M, Hurek T (2015) Roots shaping their microbiome: global hotspots for microbial activity. Annu Rev Phytopathol 53:403–424. https://doi.org/10.1146/annurev-phyto-082712-102342

Rosenblueth M, Martínez-Romero E (2006) Bacterial endophytes and their interactions with hosts. Mol Plant Microbe Interact 19(8):827–837. https://doi.org/10.1094/MPMI-19-0827

Rosser SJ, French CE, Bruce NC (2001) Engineering plants for the phytodetoxification of explosives. In Vitro Cell Dev Biol-Plant 37:330–333. https://doi.org/10.1007/s11627-001-0059-1

Ryan RP, Germaine K, Franks A, Ryan DJ, Dowling DN (2008) Bacterial endophytes: recent developments and applications. FEMS Microbiol Lett 278:1–9. https://doi.org/10.1111/j.1574-6968.2007.00918.x

Rylott EL (2014) Endophyte consortia for xenobiotic phytoremediation: the root to success? Plant and Soil 385:389–394. https://doi.org/10.1007/s11104-014-2296-1

Salt DE, Smith RD, Raskin I (1998) Phytoremediation. Annu Rev Plant Physiol Plant Mol Biol 49:643–668. https://doi.org/10.1146/annurev.arplant.49.1.643

Sandermann HJ (1994) Higher plant metabolism of xenobiotics: the 'green liver' concept. Pharmacogenet Genomics 4:225–241

Sandermann H (1999) Plant metabolism of organic xenobiotics. Status and prospects of the 'green liver' concept. In: Altman A, Ziv M, Izhar S (eds) Plant biotechnology and in vitro biology in the 21st century: Proceedings of the IXth International Congress of the International Association of Plant Tissue Culture and Biotechnology, Jerusalem, Israel, 14–19 June 1998. Springer, Dordrecht, pp 321–328. https://doi.org/10.1007/978-94-011-4661-6_74

Santoyo G, Moreno-Hagelsieb G, del Carmen Orozco-Mosqueda M, Glick BR (2016) Plant growth-promoting bacterial endophytes. Microbiol Res 183:92–99. https://doi.org/10.1016/j.micres.2015.11.008

Schäffner A, Messner B, Langebartels C, Sandermann H (2002) Genes and enzymes for *in-planta* phytoremediation of air, water and soil. Acta Biotechnol 22:141–151. https://doi.org/10.1002/1521-3846(200205)22:1/2<141::AID-ABIO141>3.0.CO;2-7

Schröder P (2007) Exploiting plant metabolism for the phytoremediation of organic xenobiotics. In: Willey N (ed) Phytoremediation: methods in biotechnology, vol 23. Humana Press, New York, pp 251–263. https://doi.org/10.1007/978-1-59745-098-0_20

Schwitzguébel J-P (2017) Phytoremediation of soils contaminated by organic compounds: hype, hope and facts. J Soil Sediment 17:1492–1502. https://doi.org/10.1007/s11368-015-1253-9

Seghers D, Wittebolle L, Top EM, Verstraete W, Siciliano SD (2004) Impact of agricultural practices on the Zea mays L. endophytic community. Appl Environ Microbiol 70:1475–1482. https://doi.org/10.1007/s11356-017-9496-1

Segura A, Ramos JL (2013) Plant–bacteria interactions in the removal of pollutants. Curr Opin Biotechnol 24:467–473. https://doi.org/10.1016/j.copbio.2012.09.011

Sessitsch A, Hardoim P, Döring J, Weilharter A, Krause A, Woyke T, Mitter B, Hauberg-Lotte L, Friedrich F, Rahalkar M, Hurek T, Sarkar A, Bodrossy L, van Overbeek L, Brar D, van Elsas JD, Reinhold-Hurek B (2011) Functional characteristics of an endophyte community colonizing rice roots as revealed by metagenomic analysis. Mol Plant Microbe Interact 25:28–36. https://doi.org/10.1094/MPMI-08-11-0204

Sessitsch A, Smalla K, Johnson L, Mitter B (2016) Editorial: special thematic issue on microbe-assisted crop production. FEMS Microbiol Ecol 92(11):fiw167. https://doi.org/10.1093/femsec/fiw167

Sessitsch A, Pfaffenbichler N, Mitter B (2019) Microbiome applications from lab to field: facing complexity. Trends Plant Sci 24:194–198. https://doi.org/10.1016/j.tplants.2018.12.004

Sheng X, Chen X, He L (2008) Characteristics of an endophytic pyrene-degrading bacterium of *Enterobacter* sp. 12J1 from *Allium macrostemon* Bunge. Int Biodeter Biodegr 62:88–95. https://doi.org/10.1016/j.ibiod.2007.12.003

Shokralla S, Spall JL, Gibson JF, Hajibabaei M (2012) Next-generation sequencing technologies for environmental DNA research. Mol Ecol 21:1794–1805. https://doi.org/10.1111/j.1365-294X.2012.05538.x

Siciliano SD, Germida JJ (1998) Mechanisms of phytoremediation: biochemical and ecological interactions between plants and bacteria. Environ Rev 6:65–79. https://doi.org/10.1139/a98-005

Siciliano SD, Fortin N, Mihoc A, Wisse G, Labelle S, Beaumier D, Ouellette D, Roy R, Whyte LG, Banks MK, Schwab P, Lee K, Greer CW (2001) Selection of specific endophytic bacterial genotypes by plants in response to soil contamination. Appl Environ Microbiol 67:2469–2475. https://doi.org/10.1128/AEM.67.6.2469-2475.2001

Siebers M, Brands M, Wewer V, Duan Y, Hölzl G, Dörmann P (2016) Lipids in plant–microbe interactions. BBA-Mol Cell Biol L 1861:1379–1395. https://doi.org/10.1016/j.bbalip.2016.02.021

Singer AC, Crowley DE, Thompson IP (2003) Secondary plant metabolites in phytoremediation and biotransformation. Trends Biotechnol 21:123–130. https://doi.org/10.1016/S0167-7799(02)00041-0

Singh OV, Jain RK (2003) Phytoremediation of toxic aromatic pollutants from soil. Appl Microbiol Biotechnol 63:128–135. https://doi.org/10.1007/s00253-003-1425-1

Smith EF (1911) Bacteria in relation to plant diseases, vol 2. Carnegie Institution of Washington, Washington

Stephenson C, Black CR (2014) One step forward, two steps back: the evolution of phytoremediation into commercial technologies. Biosci Horiz Int J Stud Res 7:1–15. https://doi.org/10.1093/biohorizons/hzu009

Strobel G, Daisy B (2003) Bioprospecting for microbial endophytes and their natural products. Microbiol Mol Biol Rev 67:491–502. https://doi.org/10.1128/MMBR.67.4.491-502.2003

Sturz AV, Christie BR, Nowak J (2000) Bacterial endophytes: potential role in developing sustainable systems of crop production. Crit Rev Plant Sci 19:1–30. https://doi.org/10.1080/07352680091139169

Sun K, Liu J, Gao Y, Jin L, Gu Y, Wang W (2014a) Isolation, plant colonization potential, and phenanthrene degradation performance of the endophytic bacterium *Pseudomonas* sp. Ph6-gfp. Sci Rep 4:5462. https://doi.org/10.1038/srep05462

Sun K, Liu J, Jin L, Gao Y (2014b) Utilizing pyrene-degrading endophytic bacteria to reduce the risk of plant pyrene contamination. Plant and Soil 374:251–262. https://doi.org/10.1007/s11104-013-1875-x

Sun K, Liu J, Gao Y, Sheng Y, Kang F, Waigi MG (2015a) Inoculating plants with the endophytic bacterium *Pseudomonas* sp. Ph6-gfp to reduce phenanthrene contamination. Environ Sci Pollut Res 22:19529–19537. https://doi.org/10.1007/s11356-015-5128-9

Sun R, Crowley DE, Wei G (2015b) Study of phenanthrene utilizing bacterial consortia associated with cowpea (*Vigna unguiculata*) root nodules. World J Microbiol Biotechnol 31:415–433. https://doi.org/10.1007/s11274-014-1796-8

Sun K, Habteselassie MY, Liu J, Li S, Gao Y (2018) Subcellular distribution and biotransformation of phenanthrene in pakchoi after inoculation with endophytic *Pseudomonas* sp. as probed using HRMS coupled with isotope-labeling. Environ Pollut 237:858–867. https://doi.org/10.1016/j.envpol.2017.11.039

Susarla S, Medina VF, McCutcheon SC (2002) Phytoremediation: an ecological solution to organic chemical contamination. Ecol Eng 18:647–658. https://doi.org/10.1016/S0925-8574(02)00026-5

Sutton NB, Maphosa F, Morillo JA, Abu AS, Langenhoff AA, Grotenhuis T, Rijnaarts HH, Smidt H (2013) Impact of long-term diesel contamination on soil microbial community structure. Appl Environ Microbiol 79:619. https://doi.org/10.1128/AEM.02747-12

Taghavi S, van der Lelie D (2013) Growth-promoting endophytic bacterium *Enterobacter* sp. 638. In: de Bruijn FJ (ed) Molecular microbial ecology of the rhizosphere. Wiley, Hoboken. pp 899–908. https://doi.org/10.1002/9781118297674.ch84

Taghavi S, Barac T, Greenberg B, Borremans B, Vangronsveld J, van der Lelie D (2005) Horizontal gene transfer to endogenous endophytic bacteria from poplar improves phytoremediation of toluene. Appl Environ Microbiol 71:8500–8505. https://doi.org/10.1128/aem.71.12.8500-8505.2005

Taghavi S, Garafola C, Monchy S, Newman L, Hoffman A, Weyens N, Barac T, Vangronsveld J, van der Lelie D (2009) Genome survey and characterization of endophytic bacteria exhibiting a beneficial effect on growth and development of poplar trees. Appl Environ Microbiol 75:748–757. https://doi.org/10.1128/AEM.02239-08

Taghavi S, van der Lelie D, Hoffman A, Zhang YB, Walla MD, Vangronsveld J, Newman LA, Monchy S (2010) Genome sequence of the plant growth promoting endophytic bacterium *Enterobacter* sp. 638. PLoS Genet 6:e1000943. https://doi.org/10.1371/journal.pgen.1000943

Tardif S, Yergeau É, Tremblay J, Legendre P, Whyte LG, Greer CW (2016) The willow microbiome is influenced by soil petroleum-hydrocarbon concentration with plant compartment-specific effects. Front Microbiol 7(1363). https://doi.org/10.3389/fmicb.2016.01363

Tervet IW, Hollis JP (1948) Bacteria in the storage organs of healthy plants. Phytopathology 38:960–967

Tétard-Jones C, Edwards R (2016) Potential roles for microbial endophytes in herbicide tolerance in plants. Pest Manag Sci 72:203–209. https://doi.org/10.1002/ps.4147

Thijs S, Sillen W, Weyens N, Vangronsveld J (2016a) Phytoremediation: state-of-the-art and a key role for the plant microbiome in future trends and research prospects. Int J Phytoremediation 19:23–38. https://doi.org/10.1080/15226514.2016.1216076

Thijs S, Sillen W, Rineau F, Weyens N, Vangronsveld J (2016b) Towards an enhanced understanding of plant–microbiome interactions to improve phytoremediation: engineering the metaorganism. Front Microbiol 7. https://doi.org/10.3389/fmicb.2016.00341

Thijs S, Weyens N, Gkorezis P, Vangronsveld J (2018) Plant-endophyte partnerships to assist petroleum hydrocarbon remediation. In: Steffan R (ed) Consequences of microbial interactions with hydrocarbons, oils, and lipids: biodegradation and bioremediation. Handbook of hydrocarbon and lipid microbiology. Springer, Cham, pp 1–34. https://doi.org/10.1007/978-3-319-44535-9_9-1

Tripathi V, Edrisi SA, Chen B, Gupta VK, Vilu R, Gathergood N, Abhilash PC (2017) Biotechnological advances for restoring degraded land for sustainable development. Trends Biotechnol 35:847–859. https://doi.org/10.1016/j.tibtech.2017.05.001

Uhlik O, Leewis MC, Strejcek M, Musilova L, Mackova M, Leigh MB, Macek T (2013) Stable isotope probing in the metagenomics era: a bridge towards improved bioremediation. Biotechnol Adv 31:154–165. https://doi.org/10.1016/j.biotechadv.2012.09.003

van Aken B (2008) Transgenic plants for phytoremediation: helping nature to clean up environmental pollution. Trends Biotechnol 26:225–227. https://doi.org/10.1016/j.tibtech.2008.02.001

van Aken B (2009) Transgenic plants for enhanced phytoremediation of toxic explosives. Curr Opin Biotechnol 20:231–236. https://doi.org/10.1016/j.copbio.2009.01.011

van Aken B (2011) Transgenic plants and associated bacteria for phytoremediation of organic pollutants. In: Moo-Young M (ed) Comprehensive biotechnology, 2nd edn. Academic Press, Burlington, pp 223–237. https://doi.org/10.1016/B978-0-08-088504-9.00330-5

van Aken B, Correa PA, Schnoor JL (2010) Phytoremediation of polychlorinated biphenyls: new trends and promises. Environ Sci Technol 44:2767–2776. https://doi.org/10.1021/es902514d

van Aken B, Tehrani R, Schnoor JL (2011) Endophyte-assisted phytoremediation of explosives in poplar trees by *Methylobacterium populi* BJ001T. In: Pirttilä AM, Frank AC (eds) Endophytes of forest trees: biology and applications. Springer, Dordrecht, pp 217–234. https://doi.org/10.1007/978-94-007-1599-8_14

van der Lelie D, Barac T, Taghavi S, Vangronsveld J (2005) Response to Newman: new uses of endophytic bacteria to improve phytoremediation. Trends Biotechnol 23:8–9. https://doi.org/10.1016/j.tibtech.2004.11.008

van der Lelie D, Taghavi S, Monchy S, Schwender J, Miller L, Ferrieri R, Rogers A, Wu X, Zhu W, Weyens N, Vangronsveld J, Newman L (2009) Poplar and its bacterial endophytes: coexistence and harmony. Crit Rev Plant Sci 28:346–358. https://doi.org/10.1080/07352680903241204

Vejan P, Abdullah R, Khadiran T, Ismail S, Nasrulhaq Boyce A (2016) Role of plant growth promoting rhizobacteria in agricultural sustainability – a review. Molecules 21:573. https://doi.org/10.3390/molecules21050573

Vishal K, Mehak B, Singh PK, Pratyoosh S (2016) Recent developments in systems biology and metabolic engineering of plant–microbe interactions. Front Plant Sci 7. https://doi.org/10.3389/fpls.2016.01421

Waigi MG, Sun K, Gao Y (2017) Sphingomonads in microbe-assisted phytoremediation: tackling soil pollution. Trends Biotechnol 35:883–899. https://doi.org/10.1016/j.tibtech.2017.06.014

Wang Y, Xiao M, Geng X, Liu J, Chen J (2007) Horizontal transfer of genetic determinants for degradation of phenol between the bacteria living in plant and its rhizosphere. Appl Microbiol Biotechnol 77:733–739. https://doi.org/10.1007/s00253-007-1187-2

Wang Y, Li H, Zhao W, He X, Chen J, Geng X, Xiao M (2010) Induction of toluene degradation and growth promotion in corn and wheat by horizontal gene transfer within endophytic bacteria. Soil Biol Biochem 42:1051–1057. https://doi.org/10.1016/j.soilbio.2010.03.002

Wang J, Zhang X, Ling W, Liu R, Liu J, Kang F, Gao Y (2017a) Contamination and health risk assessment of PAHs in soils and crops in industrial areas of the Yangtze River Delta region, China. Chemosphere 168:976–987. https://doi.org/10.1016/j.chemosphere.2016.10.113

Wang J, Liu J, Ling W, Huang Q, Gao Y (2017b) Composite of PAH-degrading endophytic bacteria reduces contamination and health risks caused by PAHs in vegetables. Sci Total Environ 598:471–478. https://doi.org/10.1016/j.scitotenv.2017.04.126

Weilharter A, Mitter B, Shin MV, Chain PSG, Nowak J, Sessitsch A (2011) Complete genome sequence of the plant growth-promoting endophyte *Burkholderia phytofirmans* strain PsJN. J Bacteriol 193:3383–3384. https://doi.org/10.1128/JB.05055-11

Weir E, Doty S (2016) Social acceptability of phytoremediation: the role of risk and values. Int J Phytoremediation 18:1029–1036. https://doi.org/10.1080/15226514.2016.1183571

Weyens N, van der Lelie D, Artois T, Smeets K, Taghavi S, Newman L, Carleer R, Vangronsveld J (2009a) Bioaugmentation with engineered endophytic bacteria improves contaminant fate in phytoremediation. Environ Sci Technol 43:9413–9418. https://doi.org/10.1021/es901997z

Weyens N, van der Lelie D, Taghavi S, Newman L, Vangronsveld J (2009b) Exploiting plant–microbe partnerships to improve biomass production and remediation. Trends Biotechnol 27:591–598. https://doi.org/10.1016/j.tibtech.2009.07.006

Weyens N, van der Lelie D, Taghavi S, Vangronsveld J (2009c) Phytoremediation: plant–endophyte partnerships take the challenge. Curr Opin Biotechnol 20:248–254. https://doi.org/10.1016/j.copbio.2009.02.012

Weyens N, Croes S, Dupae J, Newman L, van der Lelie D, Carleer R, Vangronsveld J (2010a) Endophytic bacteria improve phytoremediation of Ni and TCE co-contamination. Environ Pollut 158:2422–2427. https://doi.org/10.1016/j.envpol.2010.04.004

Weyens N, Truyens S, Dupae J, Newman L, Taghavi S, van der Lelie D, Carleer R, Vangronsveld J (2010b) Potential of the TCE-degrading endophyte *Pseudomonas putida* W619-TCE to improve plant growth and reduce TCE phytotoxicity and evapotranspiration in poplar cuttings. Environ Pollut 158:2915–2919. https://doi.org/10.1016/j.envpol.2010.06.004

Weyens N, Boulet J, Adriaensen D, Timmermans J-P, Prinsen E, Sandra VO, D'Haen J, Smeets K, van der Lelie D, Taghavi S, Vangronsveld J (2012) Contrasting colonization and plant growth promoting capacity between wild type and a gfp-derivative of the endophyte *Pseudomonas putida* W619 in hybrid poplar. Plant and Soil 356:217–230. https://doi.org/10.1007/s11104-011-0831-x

Witters N, Mendelsohn R, Van Passel S, Van Slycken S, Weyens N, Schreurs E, Meers E, Tack F, Vanheusden B, Vangronsveld J (2012a) Phytoremediation, a sustainable remediation technology? II: economic assessment of CO_2 abatement through the use of phytoremediation crops for renewable energy production. Biomass Bioenergy 39:470–477. https://doi.org/10.1016/j.biombioe.2011.11.017

Witters N, Mendelsohn RO, Van Slycken S, Weyens N, Schreurs E, Meers E, Tack F, Carleer R, Vangronsveld J (2012b) Phytoremediation, a sustainable remediation technology? Conclusions from a case study. I: energy production and carbon dioxide abatement. Biomass Bioenergy 39:454–469. https://doi.org/10.1016/j.biombioe.2011.08.016

World Health Organization (2018) Persistent organic pollutants (POPs) WHO Department of Food Safety and Zoonoses (FOS). http://www.who.int/foodsafety/areas_work/chemical-risks/pops/en/. Accessed 28 Nov 2018

Wu CH, Bernard SM, Andersen GL, Chen W (2009) Developing microbe-plant interactions for applications in plant-growth promotion and disease control, production of useful compounds, remediation and carbon sequestration. J Microbial Biotechnol 2:428–440. https://doi.org/10.1111/j.1751-7915.2009.00109.x

Yousaf S, Andria V, Reichenauer TG, Smalla K, Sessitsch A (2010a) Phylogenetic and functional diversity of alkane degrading bacteria associated with Italian ryegrass (*Lolium multiflorum*) and Birdsfoot trefoil (*Lotus corniculatus*) in a petroleum oil-contaminated environment. J Hazard Mater 184:523–532. https://doi.org/10.1016/j.jhazmat.2010.08.067

Yousaf S, Ripka K, Reichenauer TG, Andria V, Afzal M, Sessitsch A (2010b) Hydrocarbon degradation and plant colonization by selected bacterial strains isolated from Italian ryegrass and birdsfoot trefoil. J Appl Microbiol 109:1389–1401. https://doi.org/10.1111/j.1365-2672.2010.04768.x

Yousaf S, Afzal M, Reichenauer TG, Brady CL, Sessitsch A (2011) Hydrocarbon degradation, plant colonization and gene expression of alkane degradation genes by endophytic *Enterobacter ludwigii* strains. Environ Pollut 159:2675–2683. https://doi.org/10.1016/j.envpol.2011.05.031

Zhang C, Feng Y, Liu Y-W, Chang H-Q, Li Z-J, Xue J-M (2017) Uptake and translocation of organic pollutants in plants: a review. J Integr Agric 16:1659–1668. https://doi.org/10.1016/S2095-3119(16)61590-3

Zhu X, Ni X, Liu J, Gao Y (2014) Application of endophytic bacteria to reduce persistent organic pollutants contamination in plants. CLEAN-Soil Air Water 42:306–310. https://doi.org/10.1002/clen.201200314

Zhu X, Jin L, Sun K, Li S, Ling W, Li X (2016a) Potential of endophytic bacterium *Paenibacillus* sp. PHE-3 isolated from *Plantago asiatica* L. for reduction of PAH contamination in plant tissues. Int J Environ Res Public Health 13:633. https://doi.org/10.3390/ijerph13070633

Zhu X, Jin L, Kai S, Li S, Li X, Ling W (2016b) Phenanthrene and pyrene modify the composition and structure of the cultivable endophytic bacterial community in ryegrass (*Lolium multiflorum* Lam). Int J Environ Res Public Health 13:1081. https://doi.org/10.3390/ijerph13111081

Zhu X, Wang W, Sun K, Lin X, Li S, Waigi MG, Ling W (2017a) Inoculating wheat (*Triticum aestivum* L.) with the endophytic bacterium *Serratia* sp. PW7 to reduce pyrene contamination. Int J Phytoremediation 19:718–724. https://doi.org/10.1080/15226514.2017.1284747

Zhu X, Wang W, Crowley DE, Sun K, Hao S, Waigi MG, Gao Y (2017b) The endophytic bacterium *Serratia* sp. PW7 degrades pyrene in wheat. Environ Sci Pollut Res 24:6648–6656. https://doi.org/10.1007/s11356-016-8345-y

Zinniel DK, Lambrecht P, Harris NB, Feng Z, Kuczmarski D, Higley P, Ishimaru CA, Arunakumari A, Barletta RG, Vidaver AK (2002) Isolation and characterization of endophytic colonizing bacteria from agronomic crops and prairie plants. Appl Environ Microbiol 68:2198–2208. https://doi.org/10.1128/AEM.68.5.2198-2208.2002

Interaction of Copper-Based Nanoparticles to Soil, Terrestrial, and Aquatic Systems: Critical Review of the State of the Science and Future Perspectives

Vishnu Rajput, Tatiana Minkina, Bilal Ahmed, Svetlana Sushkova, Ritu Singh, Mikhail Soldatov, Bertrand Laratte, Alexey Fedorenko, Saglara Mandzhieva, Eliza Blicharska, Javed Musarrat, Quaiser Saquib, Jolanta Flieger, and Andrey Gorovtsov

Contents

V. Rajput (✉) · T. Minkina · S. Sushkova · A. Fedorenko · S. Mandzhieva · A. Gorovtsov
Academy of Biology and Biotechnology, Southern Federal University, Rostov-on-Don, Russia
e-mail: rajput.vishnu@gmail.com; tminkina@mail.ru; terra_rossa@mail.ru; afedorenko@mail.ru; msaglara@mail.ru; gorovtsov@gmail.com

B. Ahmed · J. Musarrat
Department of Microbiology, Faculty of Agricultural Sciences, Aligarh Muslim University, Aligarh, Uttar Pradesh, India
e-mail: bilalahmed.amu@gmail.com; musarratj1@yahoo.com

R. Singh
Department of Environmental Science, School of Earth Sciences, Central University of Rajasthan, Ajmer, Rajasthan, India
e-mail: ritu_ens@curaj.ac.in

M. Soldatov
The Smart Materials Research Center, Southern Federal University, Rostov-on-Don, Russia
e-mail: mikhail.soldatov@gmail.com

B. Laratte
Départment de Conception, Industrialisation, Risque, Décision, Ecole Nationale Supérieure d'Arts et Métiers, Paris, France
e-mail: bertrand.laratte@ensam.eu

E. Blicharska · J. Flieger
Department of Analytical Chemistry, Medical University of Lublin, Lublin, Poland
e-mail: eb.blicharska@gmail.com; j.flieger@umlub.pl

Q. Saquib
Zoology Department, College of Sciences, King Saud University, Riyadh, Saudi Arabia
e-mail: quaiser.saquib0@gmail.com

© Springer Nature Switzerland AG 2019

P. de Voogt (ed.), *Reviews of Environmental Contamination and Toxicology, Volume 252*, Reviews of Environmental Contamination and Toxicology Volume 252,
https://doi.org/10.1007/398_2019_34

1 Introduction

In recent years, potential effects of engineered nanoparticles (ENPs), and more so of metallic and metal oxide NPs, on aquatic and terrestrial systems have received increased attention due to their wide applications and consequential release into the environment. Metallic NPs possess unique properties for potential use in the rapidly growing nanotechnology industry (Ali et al. 2015; Arruda et al. 2015; Saleem et al. 2017). Various products containing NPs are currently in the marketplace, and many are still being added to the list (Ahmed et al. 2018b; Rajput et al. 2018c; Vance et al. 2015). The Global Market for Metal Oxide Nanoparticles indicates that the metal oxide NP production could increase from 0.27 million tons (2012) to 1.663 million tons by 2020 (The Global Market for Metal Oxide Nanoparticles to 2020). Among them, Cu-based NPs have wide applications in the field of metallurgy, electronics, automotive, fuel, transportation, machinery, etc. The annual production of Cu was approximately 18.7 million metric tons in 2015 (Keller et al. 2017), out of which a small fraction of approximately 200 tons was comprised of Cu-based NPs (Keller and Lazareva 2013). Since then, the use of Cu-based NPs has been rapidly escalating into applications such as solar cells, sensor development, catalysts, hydrogen production, drug delivery, catalysts for typical C-N cross-coupling reactions, and light-emitting diodes (Keller et al. 2017; Rajput et al. 2017b). Due to their antimicrobial and antifungal properties, Cu-based NPs are suitable for biomedical applications and are also used in water treatment (Ben-Sasson et al. 2016), textile industries (Sedighi and Montazer 2016), food preservation, and agricultural practices (Montes et al. 2016; Ponmurugan et al. 2016; Ray et al. 2015). The rapid production and multifarious applications of Cu-based NPs in various industries have necessitated the assessment of their impacts on the environment (Ahmed et al. 2018b, c).

Copper (Cu) is a naturally occurring ubiquitous element present in the environment with a concentration around 60 g per ton in the Earth's crust (Ojha et al. 2017) and essential micronutrient for plant growth at certain concentrations and is known to play important roles in mitochondrial respiration, hormone signalling, cell wall metabolism, iron mobilization, and electron transport (Yruela 2009). However, at higher concentrations, Cu is generally toxic to plants and other organisms including algae, mussels, crustaceans, and fish (Aruoja et al. 2009; Braz-Mota et al. 2018; Katsumiti et al. 2018; Ruiz et al. 2015). While there is no data available on the concentration of CuO-NPs in the soil, total Cu could range from 2 to 100 mg kg^{-1} in unpolluted soils (Nagajyoti et al. 2010). Soil receives Cu-based NPs from direct application of agricultural nano-products and industrial wastes (Adeleye et al. 2016; Rajput et al. 2017a, 2018b). The toxic action of pesticides specifically Cu-based NPs and Cu-based nano-pesticides (e.g., Kocide 3000) makes them appropriate to be used for the control of plant pathogens and pests (Anjum et al. 2015; Shahid and Khan 2017). Cu-based fungicides have been used for more than a century contributing to soil contamination based on their Cu^{2+} content, allowing them to function as a reducing or oxidizing agent in biochemical reactions. Terrestrial species can have more interactions with NPs because up to 28% of the total NPs production ends into soils (Keller and Lazareva 2013). Substantially increased production of Cu-based NPs in the last decade emphasizes the need of thorough and systematic investigation of nano-Cu release, environmental fate, bioavailability, dissolution of Cu^{+}/Cu^{2+} ions from Cu-based NPs, exposure routes, and their toxic impacts on nontarget organisms (Keller et al. 2017).

Plants are one of the most important entities and provide a very large surface area for NPs exposure via roots and aboveground parts (Dietz and Herth 2011). For instance, the air-dispersed NPs may penetrate and transport via the stomatal openings (Pullagurala et al. 2018; Raliya et al. 2016). Different plants exhibit specific behaviors toward excess metal present in the growth medium. In particular, metal-tolerant plants could limit the uptake of NPs into photosynthetic tissues by restricting the transport of metals across the root endodermis and storing them in the root cortex; hyperaccumulating plants could compile excess NPs in the harvestable tissues (Manceau et al. 2008). The exact mechanism of plant defense toward NPs toxicity is not fully understood.

At present, inadequate information is available on how Cu-based NPs affect the soil organisms, for instance, agriculturally important microbes, fungi, nematodes, and earthworms. The NPs may affect soil flora directly by inducing changes in the bioavailability of other toxins and nutrients or indirectly via interactions with natural organic compounds possible interactions with toxic organic compounds which may increase or decrease the toxicity of NPs (Haris and Ahmad 2017).

In order to get more in-depth knowledge of Cu-based NPs, this review critically assessed the literature data present over effects of Cu-based NPs on terrestrial and aquatic ecosystems, the interaction of soil microbial communities with Cu-based NPs, the bioaccumulation of Cu-based NPs in plants and their toxicity mechanism, and their biotransformation in soil (Fig. 1).

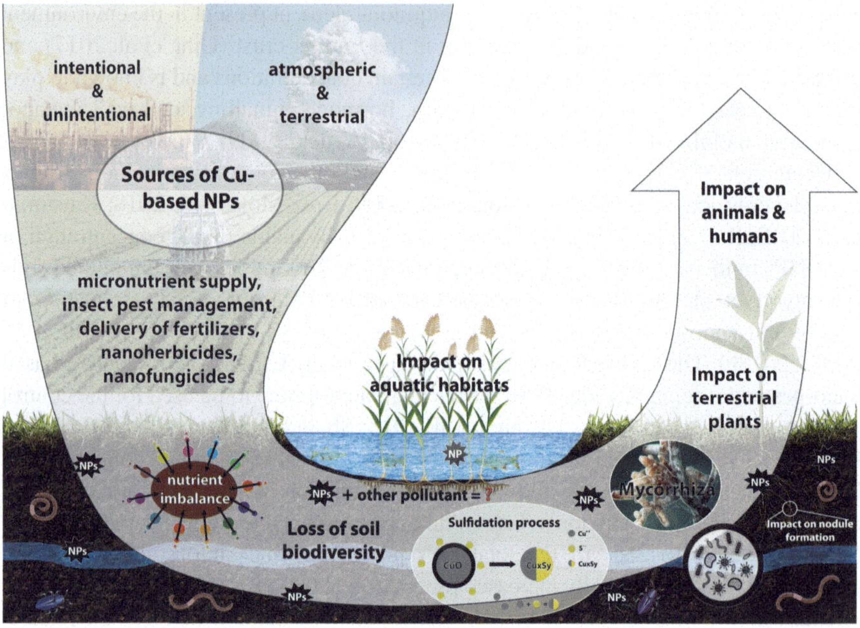

Fig. 1 Schematic of Cu-based NPs sources to environment and their effects on different ecosystems

2 Sources, Variants, and Fate of Cu-Based NPs in the Environment

Owing to diverse applications of Cu-based NPs in the nanotechnology industry, the release of nanoscale Cu in a different sphere of the environment is expected (Qiu and Smolders 2017). Sources of NPs include both the point and nonpoint sources. Point sources are comprised of production and storage units, research laboratories, disposal of nanomaterial-containing consumer products, wastewater treatment plants, etc., whereas Cu discharge through nonpoint sources occurs through wear and tear of Cu-based NPs-containing paints, cosmetic products, and cleaning agents (Rajput et al. 2018b). The Cu-NPs have potential to enter water, soil, and sediments during and at the end of their life cycle (Keller et al. 2013; Slotte and Zevenhoven 2017). Soil can receive NPs through various channels, for example, agricultural amendments of sewage sludge, atmospheric deposition, landfills, or accidental spills during industrial production (Simonin and Richaume 2015). The Cu-based NPs are available with various morphologies like Cu, CuO, Cu_2O, and Cu_3N exhibiting various oxidation states, for instance, Cu^0, Cu^I, Cu^{II}, and Cu^{III} and Cu^+ (Cu^2O) or Cu^{2+} (CuO) (Ojha et al. 2017). In soil, nanoscale Cu might be present in various forms like complexes with soil organic matters such as natural organic matter, humic acid, fulvic acid, etc. and Cu-NPs-containing pesticides including Kocide 3000 [nCu $(OH)_2$], as complex with other metal components/plant exudates, etc. (Conway et al. 2015; Gao et al. 2018; Peng et al. 2017; Servin et al. 2017a).

Due to their high density, Cu-NPs tend to settle rapidly from nano- to microscale. The Cu-NPs, both in the presence and absence of organisms, may undergo microscale aggregation with high polydispersity in water and simple salt solutions (Adeleye et al. 2014; Conway et al. 2015; Griffitt et al. 2007). In a study by Adeleye et al. (2014), only 20% Cu-NPs was detected after 6 h at pH 7.0 in NaCl (10 mM) which suggested rapid aggregation of Cu-NPs leading to sedimentation. On the other hand, natural organic matter released in the environment may reduce the Cu-NPs sedimentation; for instance, approximately 40% of Cu-NPs remained stabilized by organic matter released by fish even after 48 h (Griffitt et al. 2007). Indeed, the dissolution of CuO-NPs in aqueous medium is too slow; so much so that a within concentration range of 0.01–10 mg L^{-1}, CuO-NPs showed as little as $\leq 1\%$ dissolution after weeks in freshwater and after a month in seawater (Adeleye et al. 2014; Atha et al. 2012; Buffet et al. 2013; Conway et al. 2015; Hanna et al. 2013). A month after soil contaminated by CuO-NPs, an increase in labile fraction of the Cu was noted, which had negative effects on the *T. aestivum* growth (Gao et al. 2018).

Thus, once entered into the environment, nanoscale Cu is expected to undergo a series of transformations and partitioning that ultimately decides its fate and bioavailability to organisms.

3 Biotransformation of Cu-Based NPs in Soil

Being a less dynamic component of the biosphere, the soil system has a relatively high potential for pollutant accumulation in comparison to the atmosphere and hydrosphere. Soil not only acts as a depot for pollutants but also serves as a source of contaminant input into food chains. Additionally, the soil matrix is considered abundant in natural occurring NPs which exist in both forms: as primary particles and as agglomerates/aggregates. The natural organic matter of soil influences the bioavailability of NPs through a variety of mechanisms like electrostatic interactions, ligand exchange, hydrophobic effect, hydrogen bonding, and complexation (Philippe and Schaumann 2014). The various soil processes such as homo-/hetero-aggregation, oxidation, dissolution, sulfidation, and sedimentation may impact NPs toxicity (Adeleye et al. 2016; Conway et al. 2015; Garner and Keller 2014; Lowry et al. 2012; Miao et al. 2015; Torres-Duarte et al. 2016). Aggregation and dissolution of NPs are generally influenced by a range of environmental factors such as pH, organic matter, ionic species, and colloids. A passivation process frequently occurring under various environmental conditions is the sulfidation of CuO-NPs (Gogos et al. 2017; Ma et al. 2014). This process is expected to alter the speciation and properties of CuO-NPs significantly and might increase its apparent solubility resulting in increased bioavailability and thus eco-toxicity attributed to toxic Cu^{2+} (Ma et al. 2014).

Additionally, colloidal stability of particle is one of the critical factors controlling their fate and effects (Lowry et al. 2012). The toxicity and bioavailability of Cu change according to the Cu speciation including ionic Cu, Cu-NPs, complexed Cu, bulk Cu, oxidation states, and environmental factors such as pH, soil, water, sedimentation, organic matter, redox potential, plant species, and growth phase (Cornelis et al. 2014; Garner and Keller 2014; Zhang et al. 2018).

In soil, NPs either interact with each other forming homoaggregates or interact with different NPs and natural colloids forming hetero-aggregates (Cornelis et al. 2014; del Real et al. 2018). The process of NPs aggregation mainly impacts their colloidal stability which is among the key factors controlling NPs' fate and impact (Bundschuh et al. 2018). The extent of aggregation correlates well with the ionic strength of the medium but not with the sedimentation rate (Conway et al. 2015). The major controlling factor for Cu-based NPs sedimentation includes phosphate and carbonate content in the matrix and the oxidation state of Cu. The dissolution of Cu-based NPs is majorly hindered by sulfidation which is often regarded as passivation process for Cu/CuO-NPs. It increases the solubility of Cu/CuO-NPs resulting in enhanced bioavailability and toxicity (Ma et al. 2014). The transformation of Cu-based NPs is further influenced by geochemical properties of soil. In line with this, low translocation of Cu-NPs was observed in organic-rich soil, whereas high translocation was noticed in sandy clay soil. The highest rate for transformation to Cu ions and adsorption complexes was detected in acidic soils (Shah et al. 2016). Under slightly acidic conditions, CuO-NPs may combine with the hydrogen ions of soil and release Cu^{2+} or $Cu(OH)^{+}$. Under long-term exposure, CuO-NPs and Cu in combination with humic acid get transformed to Cu_2S and Cu-goethite complex (Peng et al. 2017).

Moreover, Wang et al. (2013) investigated the transformations of CuO-NPs in biological and environmental media and their effect over Cu-bioavailability, redox activity, and toxicity. The authors revealed that CuO-NPs underwent sulfidation process via sequential dissolution and reprecipitation mechanism to generate complex secondary aggregates of copper sulfide (CuS) NPs which are considered as active catalysts for bisulfide oxidation. Although the sulfidation is considered as a natural detoxification mechanism for heavy metals, the authors suggested that it may not permanently detoxify copper as CuS-NPs but also show redox activity through the release of Cu(I) or Cu(II) by H_2O_2 oxidation. In another study, wheat crop was exposed to CuO-NPs in a sand growth matrix, and similar transformation of CuO to Cu (I)-sulfur complexes was noticed (Dimkpa et al. 2012). Significant reduction of CuO-NPs to Cu_2S and Cu_2O was also shown in maize during root-shoot-root translocation of CuO-NPs (Wang et al. 2012). The reason behind the transformation of Cu(II) to Cu(I) in plants may be ascribed to the presence of reducing sugars which get transported from leaf cells to roots (Huang et al. 2017; Servin et al. 2017a).

The leaching and mobilization of nano-Cu ions from the source material followed by their complexation with humic acids or organic acids when secreted by fungi and contained in the plant root exudates influence the biotransformation. Although CuO-NPs are often considered as insoluble materials, the presence of organic acids such as citric and oxalic acid in the environment enhances the dissolution of

Cu and CuO-NPs which in turn increases their mobility and bioavailability to plants and animals. In addition, the nature of the organic acids also affects NPs dissolution significantly (Mudunkotuwa et al. 2012). Other factors affecting NPs dissolution include pH, dissolved organic matter, biomolecular ligands, ionic strength, etc. (Yu et al. 2018). All these factors determine the toxicity of Cu-based NPs by influencing the total dissolved concentration of Cu in the concerned media. Among these factors, the pH has an inverse relationship with dissolution. The CuO-NPs have good solubility at lower pH which is turned down as the pH increases. However, the presence of ligands including those with amine functional groups induces solubility of CuO-NPs at neutral pH (Wang et al. 2013). Recently, Kovacec et al. (2017) investigated potential efficacy of two phytopathogenic fungi, namely, *Botrytis cinerea* and *Alternaria alternata*, for biotransformation of Cu^{2+} ions, micro and nanoparticulate forms of Cu and CuO. The study revealed that *B. cinerea* could transform micro and nanoparticulate forms of Cu and CuO into Cu-oxalate complex.

Furthermore, the waterlogged conditions as in the case of paddy fields may influence NPs dissolution, mobility, bioavailability, accumulation, translocation, and transformation. Peng et al. (2017) studied bioavailability and speciation of CuO-NPs in the paddy soil and transformation of CuO-NPs in the soil-rice system. Experimental findings showed that CuO-NPs significantly reduce the redox potential of the soil and alleviate the electrical conductivity at the maturation stage of paddy. The bioavailability of CuO-NPs showed a declining trend with rice growth, but an increase was noticed after drying-wetting cycles. Most of the Cu present in the root, shoot, and leaves of the plant was found in the form of Cu-citrate. Nearly one-third of the Cu(II) was transformed to Cu(I)-cysteine, while 15.7% was present as Cu_2O in roots and 19% as Cu(I)-acetate in shoot section. In chaff, about 30% of Cu was found as Cu-citrate and Cu(I)-acetate, but no CuO was reported to reach polished rice. In another study, a higher content of Cu in the form of Cu(I) in rice grain was found in the presence of sulfur (Sun et al. 2017). It was suggested that sulfur fertilization decreases the Cu content in the root, leaf, and husk of the plant yielding higher biomass but showed higher amounts of Cu in rice grains in the form of Cu(I)-cysteine and Cu(I)-acetate.

Therefore, the mechanism of biotransformation of Cu-based NPs includes series of chemical and biochemical reactions with soil components and living organisms.

4 Interaction of Cu-Based NPs with Soil Organisms

Deliberate administration of NPs into soils might have a significant impact on the living entities, as they are extremely resistant to degradation and have the potential to accumulate in the soil. The effect of NPs may also vary with varying concentration, soil properties, and enzymatic activity. Soil properties, such as pH, texture, structure, and organic matter content, influence the structure of soil microbial community and the ability of pollutants to exert toxic effects on microorganisms (Simonin and

Richaume 2015). As NPs have the ability to mobilize soil pollutants, comparison of the toxicity of the NPs in various soil types is much required. In order to understand the influence of soil physicochemical properties on Cu-based NPs toxicity, a number of predictive models have been developed; however, these models are not always effective for other region soils (Duan et al. 2016).

The toxic effect of Cu-based NPs has been shown for beneficial soil microbes such as nitrifying bacteria, nitrogen-fixing bacteria, *Arbuscular mycorrhiza*, and other *Rhizobacteria*; however, it also influences other microorganisms. You et al. (2017) suggested that the soil types could play an important role in determining NPs toxicity over soil bacterial community composition and size. Recent studies showed that NPs might affect enzymatic and metabolic activities, nitrification potential, colony count, and abundance of soil bacterial diversity (Colman et al. 2013; Ge et al. 2011; He et al. 2016).

Copper ions released from the Cu-NPs can be toxic to both the pathogenic and beneficial bacteria (Lofts et al. 2013). The study conducted on CuO-NPs toxicity to *Saccharomyces cerevisiae* showed increased toxicity over time due to increased dissolution of Cu ions from CuO (Kasemets et al. 2009). Furthermore, Concha-Guerrero et al. (2014) have shown that CuO-NPs were very toxic for native soil bacteria, as the formation of cavities, holes, membrane degradation, blebs, cellular collapse, and lysis in the cells of soil bacterial isolates were observed. Pradhan et al. (2011) investigated the effect of CuO-NPs on leaf microbial decomposition and found a decrease in leaf decomposition rate. The bacteria from *Sphingomonas* genus and *Rhizobiales* known for their importance in remediation and symbiotic nitrogen fixation appeared susceptible to Cu-NPs (Shah et al. 2016). The NPs also have significant effects on enzymatic activities (invertase, urease, catalase, and phosphatase, dehydrogenase), microbial community structure, bacterial diversity, nutrient cycling, changes in humic substances, and biological nitrogen fixation. The CuO-NPs at 30–60 mg L^{-1} affected the microbial enzymatic activity of activated sludge (Wang et al. 2017). Several other studies also report Cu-NPs effects on soil microbial community, enzymatic activities, and reduced C and N biomass (Ben-Moshe et al. 2013; Kumar et al. 2012; Xu et al. 2015). However, the effect of Cu-based NPs on the soil microbial community has rarely been explored. While Cu-based NPs are known to exhibit antimicrobial properties (Ingle et al. 2014), it is necessary to observe their impact on symbiotic microorganisms. It can be assumed that NPs, besides influencing plant and microbes, could affect plant-microbe associations either directly or indirectly. In this context, one of the classical examples is mycorrhizal symbiosis, which promotes plant growth enhancing the plant nutrient acquisition through uptake of mineral nutrients. The formation of Cu-NPs at the soil-root interface with the assistance of endomycorrhizal fungi was shown in *Phragmites australis* and *Iris pseudacorus*, and this mechanism helped to alleviate metal stress (Manceau et al. 2008). On the other hand, metallic NPs were shown to inhibit mycorrhizal plant growth (Feng et al. 2013).

Furthermore, the CuO-NPs induced morphological and genetic alterations in leaf litter decomposing fungus which could impact organic matter decomposition rate (Pradhan et al. 2011). A significant negative impact on bacterial hydrolytic activity, oxidative potential, community composition, and population size was also observed in Bet-Dagan soil (Frenk et al. 2013). Cu-based NPs have also been reported to affect the growth and functionality of green algae, cyanobacteria, and diatoms (Anyaogu et al. 2008). The most recent findings on Cu-based NPs action on bacteria are summarized in Table 1.

The findings of recent studies dealing with the NPs action on bacteria are often controversial (Table 1). Though, most studies show the increased toxicity of Cu-based NPs in comparison to ionic copper at similar dose rates (VandeVoort and Arai 2018). Interesting results were also obtained when NPs interaction with pesticides was studied. Parada et al. (2019) reported no major shift in microbial species composition; however, the degradation of the pesticide was reduced. The possible explanation for this was given by Parra et al. (2019), wherein they showed a decrease in spreading of pesticide degradation genes bearing plasmids among the bacterial community. Therefore, the current scenario demands the exploration of NPs toxicity mechanism on the soil microorganisms.

In addition, some studies report that Cu-based NPs can also have adverse effects on multicellular soil organisms. For instance, the CuO-NPs affected growth and neuron morphology of a transgenic *Caenorhabditis elegans* (Mashock et al. 2016) and disturbed immunity and reduced population density of a common earthworm *Metaphire posthuma*, which is mostly distributed across the Indian subcontinent (Gautam et al. 2018).

Considering the presence of Cu-based NPs in the soil, it is imperative to study their influence on soil biodiversity. The reviewed information indicates that NPs affected soil microbial community by decreasing their abundance, enzymatic activities, and soil microbial biomass. Therefore, the decrease in soil microbial biomass could be a sensitive indicator for microbial changes in soils.

5 Uptake and Bioaccumulation of Cu-Based NPs in Plants

The NPs are taken up by plant roots and transported to the aboveground plant tissues through the vascular system, depending on the composition, shape, size of NPs, and anatomy of the plants (Rico et al. 2011). On the other hand, some NPs remain adhered to the plant roots. It is well understood that NPs enter plant tissues either via root tissues (root tips, rhizodermis, and lateral root junctions) or the aboveground organs and tissues (cuticles, trichomes, stomata, stigma, and hydathodes) as well as through wounds and root junctions. Interestingly, in the event of NP-plant interaction, some metal-tolerant plants could limit the uptake of NPs into the photosynthetic tissues by restricting the transport of metals across the root endodermis and storing them on the root cortex, whereas hyper-accumulating plants can take up excess amounts of NPs in the harvestable tissues of plants (Manceau et al. 2008). It has been suggested that the plants can accumulate NPs in their original form or as metal ions (Cota-Ruiz et al. 2018). However, the uptake and bioaccumulation vary with varying

Table 1 The effects of Cu-based NPs on bacteria

Group of bacteria	Dose and type of the NPs	Observed effects	Reference
B. subtilis; *E.coli*	CuO-NPs (<50 nm) 0–100 mg L^{-1}	Cu-NPs were more toxic to *B. subtilis* than ionic forms of Cu. The opposite was observed for *E. coli*. The different action on Gram-positive and Gram-negative bacteria implies the possible role of NPs in changing the composition of microbial communities in the environment	Adhikari et al. (2018)
E.coli	Cu and CuO-NPs (<100 nm) 0.1–100 mg L^{-1}	The toxic effect of NPs in saline solution was much greater than in distilled water. The fresh suspension of NPs had a more pronounced toxic effect than the same suspension after 24 h. The findings highlight the role of aggregation and dispersion medium in NPs toxicity to the bacteria	Zakharova et al. (2018)
Soil nitrifying bacteria	CuO-NPs (35 nm) 1–1,000 mg L^{-1}	CuO-NPs suppressed the nitrification kinetics in batch experiments. The level of suppression had no significant difference to the respective ionic Cu concentration	VandeVoort and Arai (2018)
Soil bacterial community	Cu-NPs (40–60 nm) 500–1,500 mg kg^{-1}	The combined pollution with atrazine and Cu-NPs has led to a significant decrease in atrazine dissipation. However, the microbial communities were not affected, and the observed effect is attributed to physical-chemical interactions	Parada et al. (2019)
Soil bacterial community	Cu-NPs (40–60 nm) 10–100 mg L^{-1}	The presence of Cu-NPs hampered the conjugation and plasmid transfer in bacteria. Therefore, the application of Cu NPs to the soil can slow the degradation of pesticides in soils	Parra et al. (2019)
Soil bacteria involved in carbon and nitrogen cycles	CuO-NPs (<50 nm) 0.1–100 mg kg^{-1}	The microbial activity related to carbon and nitrogen cycling (denitrification, nitrification, and soil respiration) was significantly decreased after the addition of CuO-NPs. It was noted that plant presence (in experimental variants with wheat) did not mitigate CuO-NP toxicity for microorganisms	Simonin et al. (2018)
Soil bacterial community	CuO-NPs (240 nm) 10–1,000 mg kg^{-1}	The addition of CuO-NPs significantly reduced the microbial biomass. The adverse effect was partly mitigated in organic-rich soils	Shi et al. (2018)

(continued)

Table 1 (continued)

Group of bacteria	Dose and type of the NPs	Observed effects	Reference
Sulfate-reducing bacterium; *Desulfovibrio vulgaris*	CuO-NPs (<50 nm) 1–500 mg L^{-1}	CuO-NPs had pronounced adverse effects on the bacterium. The catabolic and anabolic activity, sulfate reduction rate was decreased, genes involved in electron transfer and respiration were downregulated. The observed effects were attributed to ROS generation	Chen et al. (2019)

physicochemical features of NPs (Ahmed et al. 2018b; Peng et al. 2015; Rico et al. 2011, 2015).

In a study, the translocation and biotransformation of CuO-NPs in rice plants were explored. It was revealed that CuO-NPs get accumulated in epidermis and exodermis regions of the plants and get precipitated with citrate or phosphate ligands or get bound to amino acids forming Cu-cysteine, Cu-citrate, and $Cu_3(PO_4)_2$ kind of products or get reduced to Cu(I) (Peng et al. 2015). Cu(I) is a highly redox active species capable of producing hydroxyl radicals by Fenton-like reactions, and so its presence in even smaller quantities has significant biological importance. Servin et al. (2017a) compared bioaccumulation of un-weathered and weathered CuO-NPs, bulk, and ions in lettuce plants after 70 days. In the case of CuO bulk, weathered material was found to decrease Cu accumulation in plant roots, whereas weathering had a positive impact on bioaccumulation of NPs. The authors further unearthed that in roots exposed to weathered NPs, the major fraction of Cu, i.e., 94.2%, was present in oxidized form as CuO, while the rest of the fraction, i.e., 5.7%, could bind to sulfur in reduced form as Cu_2S. In contrast, roots exposed to un-weathered NPs showed negligible biotransformation. As the aging and weathering have a profound effect on the particle size, particle-size distribution, surface properties, composition, reactivity, etc., it is an important aspect which needs to be considered while assessing the environmental implication of Cu-based NPs. Similarly, the translocation and biotransformation of NPs are plant-specific phenomena which require adequate attention.

The nano-phytotoxicity studies on accumulation and uptake of NPs have generated important data for understanding the fate of Cu-based NPs in plants (Ingle et al. 2014; Ma et al. 2010). Once NPs infiltrate the plant system, they may traverse to different organs (leaves, stem, and fruits) or may get compartmentalized at different locations, viz., vacuoles, walls, stellar system, cytoplasmic matrix, lipid envelopes, and nucleus (Ahmed et al. 2018b; Rajput et al. 2017b, 2018a; Rastogi et al. 2017). The translocation efficiency varies greatly in different plant species, for instance, alfalfa translocates 3–5% of Cu from root to shoot on exposure to 0–20 mg L^{-1} Cu-NPs, whereas only 0.5–0.6% translocation was observed in lettuce (Hong et al. 2016). Before the plant uptake, the dissolution of Cu-NPs increases the likelihood that Cu is internalized as Cu^{2+} ions or in the form of organic complexes (Keller et al.

2017). A recent study revealed the adsorption and accumulation of Cu-based NPs in tomato plants lead to the adsorption of nano-CuO on the roots (Ahmed et al. 2018b). Similarly, maize roots showed 3.6-fold greater Cu content under CuO-NPs treatments (Wang et al. 2012). Also, the Cu content was seven times higher in shoots of maize treated with 100 mg L^{-1} CuO-NPs. In this context, Zuverza-Mena et al. (2015) also reported the translocation of Cu-based NPs in cilantro and their significant accumulation in shoots. Differential accumulation profile of CuO-NPs has been reported in ryegrass and radish (Atha et al. 2012). Wheat and bean seedlings grown on dual agar media have been adequately discussed pertaining to the bioavailability of Cu-NPs and their relationship between accumulation and uptake (Woo-Mi et al. 2008). Cu-NPs were toxic to both plants and also bioavailable. A Cu ion released from Cu-NPs has negligible effects in the studied concentration range, and the apparent toxicity is clearly due to Cu-NPs. Bioaccumulation increased with increasing concentration of Cu-NPs and agglomeration of particles was observed in the plant cells by using transmission-electron microscopy-energy-dispersive spectroscopy (TEM-EDX). In shoots of wheat grown in the sand matrix, the bioaccumulated Cu was detected as Cu(I)S complex and CuO (Dimkpa et al. 2012). The level of Cu accumulation in wheat shoots under CuO-NPs exposure was almost equal to the concentrations quantitated in bulk (Dimkpa et al. 2012).

In a very recent study, Keller and co-workers exposed leaf tissues of lettuce, collard green, and kale to nano-CuO and detected CuO-NPs in leaf surfaces by use of single particle-inductively coupled plasma-mass spectroscopy (sp-ICP-MS) (Keller et al. 2018). Among all three vegetables, lettuce retained the highest amount of CuO-NPs on leaf surface even after washing. For this retention, the varying degrees of leaf surface roughness and hydrophilicity among the tested vegetables have been suggested to play an important role in holding CuO-NPs (Keller et al. 2018). Overall the data from these studies indicate that certain fractions of CuO-NPs are taken up by plants which may result in undesirable accumulation in edible plant tissues ultimately exposing humans via the food chain.

The bioaccumulated nano-Cu or CuO is also subject to transportation and transformation in plants (Ahmed et al. 2018c). For instance, the treatment of hydroponically cultured lettuce plant with CuO/Cu-based NPs caused a greater accumulation of Cu than cupric ions (Trujillo-Reyes et al. 2014). Additionally, the xylem- and phloem-based transport system to shoots and back to roots were proposed for CuO-NPs accumulation in root cells, cytoplasm, intracellular space, and nuclei of xylem and cortical cells. However, the CuO-NPs were reduced from Cu (II) $\rightarrow$ Cu (I) in due course of translocation (Wang et al. 2012). A similar transformation of CuO-NPs has been reported with an elevation in the degree of saturation of fatty acids (Yuan et al. 2016). In another study, when *Zea mays* were exposed to CuO-NPs, ionic, and bulk CuO, the Cu content in root and shoot of the plant was found enhanced under CuO-NPs (Wang et al. 2012). A micro X-ray fluorescence (μXRF) study revealed that Cu-NPs may get accumulated in outer parts of the root (Servin et al. 2017a). The translocation of Cu-NPs also varies depending upon the growth media. For instance, alfalfa, lettuce, and cilantro exposed to CuO, Cu, and $Cu(OH)_2$ NPs-based pesticide in soil showed >87–99%

Cu accumulation mostly in roots with very little transportation to shoots and negligible in leaves (Hong et al. 2015; Zuverza-Mena et al. 2015). In some recent studies, Cu-NPs were also detected in leaves, stems, and fruits of cucumber and tomato when grown in soil system (Zhao et al. 2016a). The uptake of CuO-NPs in tomato, alfalfa, cucumber, and radish seedlings was also noticed in the range of 4–1,748 μg g^{-1} dry biomass when grown on semisolid agar media (Ahmed et al. 2019). In a comparative study between soil and hydroponically grown tomato plants, the organ-wise distribution of CuO-NPs in soil culture was found lesser than in hydroponic (Ahmed et al. 2018b). The Cu in soil-grown root and shoot of tomato plants was found lesser by 20% and 33% than in hydroponically grown plants (Ahmed et al. 2018b). This difference could be attributed to the NPs cluster formation due to the homo-/hetero-aggregation processes of the soil system. Besides root exposure, the atmospheric presence of Cu-based NPs also triggers their bio-uptake. For instance, during the foliar applications of Cu-NPs, most of the Cu remained in fruits or leaves with a little transport via phloem to roots. For example, *Lactuca sativa* exposed to Cu-based nano-pesticide accumulated 1,350–2,010 mg Cu kg^{-1} dry biomass after 30 days (Zhao et al. 2016a, b). A small fraction (17–56 mg kg^{-1}) of Cu was also found in roots via phloem transport (Zhao et al. 2016a). In a study, the microscopic analysis showed the presence of dense material in root cells of *O. sativum* L. treated with CuO-NPs and confirmed the presence of Cu by bulk-X-ray absorption near edge structure (XANES), and interestingly the most dominant form of dense material was CuO (Peng et al. 2015).

Being very small in size, NPs have the potential to enter, translocate, and penetrate physiological barriers to travel within the plant tissues, and microscopic studies showed the accumulation of NPs in various parts of the plant (Ahmed et al. 2018a; Rajput et al. 2018a, d).

6 Toxicity of Cu-Based NPs in Plant System

The long-term effects of Cu-based NPs accumulation in plant systems are still scarcely known. It has been suggested that the Cu-based NPs may cause morphological, physiological, genetic, and epigenetic changes which may alter plant growth and nutritional status. Plants as primary producers are very critical for the sustainability of an ecosystem and functions as an indispensable link for perpetual food supply and human nutrition. In the environment, plant roots make close associations with soil particles and virtually everything that enters in the soil system (Ahmed et al. 2017; Anjum et al. 2013). Variants of Cu-based NPs once released in the environment may eventually enter either intentionally or accidentally into the soil-plant system. Plants in soil environment can be the nontarget organisms of Cu-based NPs. The critical toxicity level of Cu in many crop species varies between 20 and 30 μg g^{-1} leaf dry biomass (Anjum et al. 2015; Yruela 2009). Thus, the potential toxicity assessment of Cu-based NPs to plants is relevant to a large extent. Several studies have reported the impact of different species of Cu-NPs in various culture media such as agar, hydroponic nutrient solution, sand, filter paper, soil, and soil-sand mixtures (Dimkpa et al. 2013; Kim et al. 2013; Moon et al. 2014; Musante and White 2012) (Table 2). The exact mechanism of plant defense under NPs toxicity is not fully understood. Generally, the phytotoxicity of NPs expressed in two steps: (1) chemical toxicity based on chemical composition and (2) stress stimuli caused by the surface, size, or shape of the NPs. The antioxidant defense machinery of plants becomes activated against external/internal NPs stress stimuli. Underexposure with NPs having enough physicochemical features to exert toxicity, plants trigger their antioxidant defense mechanism to prevent oxidative damage, as well as enhance their resistance toward NPs toxicity. For instance, cucumber plants grown hydroponically in the presence of CuO-NPs (50 nm) were found with augmented antioxidative enzymes, viz., catalase (CAT), peroxidase (POD), and superoxide dismutase (SOD) (Kim et al. 2012). However, *C. sativus* when grown hydroponically in the presence of Cu-NPs (10–30 nm) experienced significant phytotoxic effects which were not ameliorated by antioxidant enzymes adequately (Mosa et al. 2018). The NPs arbitrated phytotoxicity is predominantly related to their physicochemical properties. The Cu-based NPs cause phytotoxicity via the dissolution and release of higher concentration of ions such as Cu^{2+} or the production of excess reactive oxygen species (ROS) (Ahmed et al. 2019; Letelier et al. 2010). ROS can affect mitochondrial respiration, apoptosis, and lipid peroxidation in the cell membrane and induce a range of antioxidant responses (Dimkpa et al. 2012; Shaw and Hossain 2013). Recent studies of CuO-NPs phytotoxicity showed negative impacts on seed germination and overall plant growth of various crops such as *Lactuca sativa* (100–300 mg L^{-1}), *Medicago sativa* (0–20 mg L^{-1}), *Triticum aestivum* (200 mg L^{-1}), *Vigna radiata* (500 mg L^{-1}), *Zea mays* (2–100 mg L^{-1}), *Cucumis sativus* (100–600 mg L^{-1}), *Oryza sativa* (0–1,000 mg L^{-1}), *Brassica juncea* (0–1,500 mg L^{-1}), and *Glycine max* (50–500 mg L^{-1}) (Rajput et al. 2017b).

The studies pertaining to toxicity assessment of Cu and Cu-based NPs and understanding of its molecular mechanism warrant more systematic and in-depth investigations. The available data on the toxicity, chemistry, and Cu-NPs plant interactions suggesting adverse outcomes on plant growth are presented in Table 2.

6.1 *Effects on Seed Germination, Morphometry, and Plant Growth*

Seed germination commences a plant's physiological process, and therefore it is an important attribute when toxicity of a xenobiotic is examined. The Cu-based NPs have been found to inhibit seed germination in various crops (Table 2). For instance, *Coriandrum sativum* cultivated in soil mixed with 20 and 80 mg kg^{-1} of each Cu, CuO, and $Cu(OH)_2$ NPs (Kocide and CuPRO) exhibited significant ($p \leq 0.05$) reduction in seed germination (Zuverza-Mena et al. 2015). In another study, the seed germination by CuO-NPs was reduced to almost 50%. Similarly, treatment with

Table 2 Representative studies investigating the impact of Cu-based nanoparticles on plants

Plants	Exposure media	Concentration range	Effective concentration	Toxic impact	Reference
Coriandrum sativum	Soil	0–80 mg kg^{-1}	20 kg^{-1}	~50% reduction in relative seed germination, significant accumulation of Cu in shoot, significantly reduced B, Zn, Mn, Ca, Mg, P, and S in shoots and shoot elongation by 12.4% compared with control	Zuverza-Mena et al. (2015)
Hordeum vulgare L.	½ MS medium	0.5–1.5 mM	1.5 mM	Decrease in root/shoot length and weight, ROS generation in leaves, marked reduction in root cell viability, ~1.8-fold decline in chlorophyll content, ~twofold to eightfold increase in H_2O_2 content, oxidative damage to lipid membrane, increased SOD, APX, GR enzyme activity, change in MDA/DHA ratio	Shaw et al. (2014)
Solanum lycopersicon	Hydroponic solution and soil	0.02–2 mg mL^{-1}	0.2 mg mL^{-1}	Accumulation and translocation in plan tissues, enhanced activity of antioxidant enzymes, generation of ROS, dissipation of mitochondrial membrane potential, plant cell death, macromolecular changes, and DNA damage	Ahmed et al. (2018b)
Cucumis sativus and *Zea mays*	Soil	1,000 mg kg^{-1}	1,000 mg kg^{-1}	Inhibition of plant biomass, root damage and inhibition of root elongation, reduction in soil enzyme activity	Kim et al. (2013)

(continued)

Table 2 (continued)

Plants	Exposure media	Concentration range	Effective concentration	Toxic impact	Reference
Raphanus sativus	In petri dishes	10–1,000 mg L^{-1}	1,000 mg L^{-1}	Significant accumulation of oxidatively modified, mutagenic DNA lesions and strong plant growth inhibition	Atha et al. (2012)
Phaseolus radiatus and *Triticum aestivum*	Agar media	200–1,000 mg L^{-1}	335 mg L^{-1} for mung bean and 570 mg L^{-1} for wheat	Reduction in length of seedlings	Woo-Mi et al. (2008)
Lactuca sativa L.	Hydroponic	0.1–0.8 mg L^{-1}	0.1 mg L^{-1}	Reduction in seedling growth	Liu et al. (2016)
T. aestivum	Soil	500 mg kg^{-1}	500 mg kg^{-1}	Significant decrease in root maximal length, aging increased toxicity	Gao et al. (2018)
H. vulgare	Soil	50–1,000 mg kg^{-1}	500 mg kg^{-1}	Soil aging increased toxicity and suppressed root elongation	Qiu and Smolders (2017)
Oryza sativa	In nutrient solution and soil	50–1,000 mg kg^{-1}	500 mg kg^{-1}	Significant reduction in water content of root, shoot, and grain yield	Peng et al. (2017)
C. sativus	Foliar application	50–200 mg L^{-1}	200 mg L^{-1}	Decrement in net photosynthesis and transpiration rate of leaves, reduction in fruit firmness and fruit molybdenum	Hong et al. (2016)
L. sativa	Foliar application	1,050–2,100 mg L^{-1}	1,050 mg L^{-1}	Alteration in metabolite levels, TCA cycle, and amino acid-related biological pathways were disturbed. Some antioxidant levels were significantly decreased	Zhao et al. (2016b)
C. sativus	Soil	200–800 mg kg^{-1}	400 mg kg^{-1}	Caused perturbation of iron uptake, metabolite profile of fruits was distinctively altered,	Zhao et al. (2017b)

				C- and N-related pathways specifically galactose metabolism and TCA cycle were perturbed	
C. sativus	Hydroponic	10 and 20 mg L^{-1}	10 and 20 mg L^{-1}	Interfered with uptake of a number of Na, P, S, Mo, Zn, and Fe, nCu triggered significant metabolic changes in leaves and root exudates	Zhao et al. (2016a)
Arabidopsis thaliana	Hydroponic	100 mg L^{-1}	100 mg L^{-1}	More Cu accumulation in CuO-NPs than Cu^{2+} exposure, sequestration of nCuO in root cell vacuoles and transformation of Cu(II) → Cu(I)Cl, elevation in degree of fatty acid saturation in plant cells through oxidative stress	Yuan et al. (2016)
A. thaliana	Hydroponic	20–50 mg L^{-1}	20 mg L^{-1}	Inhibition of seedling growth as well as the germination of their pollens and harvested seeds, toxicity to biomass, relative growth rate, root morphology change, upregulation of Fe-SOD and Aux/IAA gene as a result of oxidative stress and root inhibition	Wang et al. (2016)
L. sativa and *Medicago sativa*	Hydroponic	5–20 mg L^{-1}	10 mg L^{-1}	All Cu species reduced root length by 49% in both plants, decreased P and Fe in lettuce shoot, reduced catalase, and enhanced ascorbate peroxidase activity	Hong et al. (2015)

(continued)

Table 2 (continued)

Plants	Exposure media	Concentration range	Effective concentration	Toxic impact	Reference
Phaseolus vulgaris	Sand	100–500 mg kg^{-1}	500 mg kg^{-1}	Amount of Fe, Mn, Zn, and Ca were reduced, inhibited ferric reductase but stimulated cupric reductase activity	Dimkpa et al. (2015)
L. sativa	Hydroponic	10 and 20 mg L^{-1}	20 mg L^{-1}	Reduction in water content (60–61%), root length by 51%, and dry biomass (69%). Significant accumulation of Cu in roots, increased CAT activity, decrease Mn, Ca, Mg, and P content	Trujillo-Reyes et al. (2014)
O. sativa	½ MS	0.5–1.5 mM	1 mM	Reduction in seed germination, loss of root cell viability, severe oxidative burst, higher lipid peroxidation in leaves, enhanced proline accumulation, decline in carotenoids level	Shaw and Hossain (2013)
T. aestivum	Sand	500 mg kg^{-1}	500 mg kg^{-1}	Plant growth increased dissolution of CuO, marked reduction in root length	Dimkpa et al. (2012)
Allium cepa	Hydroponic	5–80 mg L^{-1}	20–80 mg L^{-1}	Surface of root cap and meristematic zone were damaged, apical meristem of roots stopped division, nucleus of meristematic cells deformed, plasmolysis occurred, cell and nuclear membrane fractured, lipid peroxidation increased, roots appeared to be corroded	Deng et al. (2016)

Nigella sativa	In petri dishes	0.25–1 μg mL^{-1}	0.5 and 1 μg mL^{-1}	Induced disturbance in mitosis and meiosis showed similar responses as that of EMS and gamma radiations in relation to physiological attributes	Kumbhakar et al. (2016)
A. cepa	Hydroponic	10–100 μg mL^{-1}	40–100 μg mL^{-1}	Decline mitotic index and increase in abnormality index, chromosomal aberrations such as sticky and disturbed chromosomes in metaphase/anaphase, c-metaphase, bridges, laggard, disturbed telophase, and vacuolated nucleus were observed	Nagaonkar et al. (2015)
T. aestivum	MS medium	15.6 mM	15.6 mM	Decreased root cell proliferation, induced cell death; however Cu-NP toxicity was lower than Cu^{2+}	Zhang et al. (2018)
Cucurbita pepo	Hydroponic	100 and 500 mg L^{-1}	100 and 500 mg L^{-1}	Reduction in photosynthetic rate, plant biomass, and transpiration volume	Musante and White (2012)
Z. mays	Hydroponic	2–100 mg L^{-1}	100 mg L^{-1}	Reduction in root length, biomass, and seedlings showed chlorotic symptoms	Wang et al. (2012)
C. sativus	In petri dishes	100–600 mg L^{-1}	600 mg L^{-1}	Inhibition of root elongation and seed germination, differential expression of 34 proteins in seeds	Moon et al. (2014)
Glycine max	1/2 MS medium	50–500 mg L^{-1}	500 mg L^{-1}	Significant reduction in shoot growth, weight, and total chlorophyll content, significant	Nair and Chung (2014)

(continued)

Table 2 (continued)

Plants	Exposure media	Concentration range	Effective concentration	Toxic impact	Reference
				increase in H_2O_2 levels, POD activity, lignification of root cells, ROS generation	
Transgenic and conventional cotton	In nutrient solution	10, 200, and 1,000 mg L^{-1}	1,000 mg L^{-1}	Inhibited the growth, development, nutrient content, and phytohormones	Van et al. (2016)
H. sativum	Hydroponic	10,000 mg L^{-1}	10,000 mg L^{-1}	Reduced root and shoot length, affected stomatal aperture and root morphology, reduced metaxylem number, affected chloroplast, and mitochondrial ultrastructure	Rajput et al. (2018a)
Origanum vulgare	Soil	0–200 mg kg^{-1}	200 mg kg^{-1}	Affected agronomical and physiological parameters decreased leaf starch, total sugar, and reducing sugar	Du et al. (2018)
Arachis hypogaea	Soil and sand	50 and 500 mg kg^{-1}	500 mg kg^{-1}	Decreased the grain weight, total amino acid content	Rui et al. (2018)
L. sativa and *Brassica oleracea*	Foliar application	0, 10, or 250 mg per plant	250 mg per plant	Deformed stomata, decreased plant weight, net photosynthesis	Xiong et al. (2017)
Chlamydomonas reinhardtii	In aquatic ecosystem	0.1–1,000 mg L^{-1}	Toxic response appears from 0.1 mg L^{-1} after 72 h of treatment. EC50 is 150.45 ± 1.17 mg L^{-1}	Decreased carotenoids levels, decreased cell metabolism activity, increased reactive species level (190 ± 0.45% at 1,000 mg L^{-1}), increased lipid peroxidation of cellular membranes (73 ± 2% at 1,000 mg L^{-1})	Melegari et al. (2013)

Lemna gibba	In freshwater	0.68, 1.04, 2.08, 4.51 g L^{-1} for CuO NPs; 0.25, 0.42, 0.72, 1.24 g L^{-1} for CS-CuO NPs	4.5 g L^{-1} for CuO NPs, 0.4 g L^{-1} for coated CuO after 48 h of exposure	Formation of reactive oxygen species, decreased growth rate	Perreault et al. (2014)
Nitellopsis obtusa and *Chlorella*	Nanosuspensions	3–100 mg L^{-1}	100 mg L^{-1}	Lethality, influence of a cell wall on nanotoxicity	Manusadzianas et al. (2012)
Pseudokirchneriella	Algal medium	0.5–6.5 mg L^{-1}	0.71 mg Cu L^{-1}	Growth inhibition	Aruoja et al. (2009)
Lemnaceae species	Suspensions	1.15, 0.84, 0.64 mg L^{-1}	1.15 mg L^{-1} for *Spirodela polyrhiza*, 0.84 mg L^{-1} for *Lemna minor*, 0.64 mg L^{-1} for *Wolffia arrhiza*	The inhibiting effect on the total frond area, the frond number, and the dry weight-based relative growth rate	Song et al. (2015a)
T. aestivum L.		15.6 μM	15.6 μM	Oxidative stress and root growth inhibition	Zhang et al. (2018)
Pistia stratiotes L.	Aquatic medium	0.75 mg L^{-1}	–0.75 mg L^{-1}	Decreased the amount of ascorbic acid, the total content of amino acids, reduction of the contents of acylcarnitines	Olkhovych et al. (2016)
Lemna minor	Modified Steinberg medium	0–200 mg L^{-1}	–50 mg L^{-1}	The changes in plant growth, chlorophyll content, enzyme activities (POD, CAT, SOD, MDA, and cell damage)	Song et al. (2016)
Elodea densa Planch	Aquatic medium	0.025–5.0 mg L^{-1}	–1.0 mg L^{-1}	Activation lipid peroxidation, increased activities of catalase and superoxide dismutase, photosynthesis suppression	Nekrasova et al. (2011)

Cu-NPs at 80 mg kg^{-1} reduces the shoot elongation by 11% (Zuverza-Mena et al. 2015). The CuO-NPs (~18.4 nm) at 0.02–2 mg mL^{-1} also cause severe toxicity in tomato plants (Ahmed et al. 2018b). Furthermore, *Solanum lycopersicon* plants are grown in both soils, and hydroponic media showed significant internalization of Cu in different plant organs with oxidative burst and reduction in plant height and weight (Ahmed et al. 2018b). Moreover, the Cu, CuO, and core-shell Cu/CuO-NPs at different concentrations caused severe reduction in root length of *Hordeum vulgare* L. (Shaw et al. 2014), *H. sativum* distichum (Rajput et al. 2018a), *H. vulgare* (Qiu and Smolders 2017), *Z. mays*, *C. sativus* (Kim et al. 2013), *T. aestivum* (Gao et al. 2018; Woo-Mi et al. 2008), and *L. sativa* (Liu et al. 2016; Trujillo-Reyes et al. 2014). The CuO-NPs (~40 nm) at 500 mg kg^{-1} soil as fresh and after 28 days of mixing of CuO-NPs with soil caused a significant decrease in maximal root length (Gao et al. 2018). In the same study, it has been suggested that the exudates secreted from wheat roots in CuO-NPs amended soil enhanced the dissolution of Cu ions in pore water, which played an important role in enhanced phytotoxicity (Gao et al. 2018). Similarly, in a study by Qiu and Smolders (2017), CuO-NPs (~34 nm) at various concentrations ranging from 50 to 1,000 mg kg^{-1} at two different pH (4.8 and 5.8) increase the toxicity of CuO-NPs affecting root elongation. The CuO-NPs inhibited *C. sativus* seed germination when administered at 600 mg L^{-1}. At this rate, only 23.3% germination was recorded over untreated of control (Moon et al. 2014). Some earlier studies also reported that CuO-NPs reduced *C. pepo* biomass by 90% (Stampoulis et al. 2009), seedling growth of *Phaseolus radiatus* and *T. aestivum* (Woo-Mi et al. 2008), shortened primary and lateral roots of the *B. juncea* L (Nair and Chung 2015a), affected agronomical/physiological parameters in *Origanum vulgare* (Du et al. 2018), and decreased root growth in *M. sativa* grown in hydroponic culture (Hong et al. 2015). In *Allium cepa*, 80 mg CuO-NPs L^{-1} damaged the root cap and meristematic zone and reduced the growth of the root tip (Deng et al. 2016).

Morphometric observations indicated a decline in root and shoot growth for Cu-based NPs-treated plants. Also, Cu-based NPs pose deleterious effects on plant germination (Deng et al. 2016; Moon et al. 2014; Nair and Chung 2015a; Rajput et al. 2018a, b). The reduction in root and shoot growth could limit the surface area for water uptake and photosynthesis, respectively, and consequently affects the plant performance.

6.2 *Effects on Cellular Ultrastructure*

Several studies on the ultrastructure of plant cells after Cu-based NPs exposure showed remarkable changes in plant roots and leaves. In roots, violations of the integrity of the cell wall of the epidermis and endoderm, vacuolization and disorganization of fragments in the endoplasmic reticulum, swelling of the mitochondria, and destruction of the mitochondrial cristae have been observed with rare leucoplasts with disorganized and partially destroyed thylakoid. In the chloroplasts of the leaf parenchyma, the size of starch grains and plastoglobules increased significantly; the area of the thylakoids decreased and inter-thylakoid space expanded (Rajput et al. 2018d). These changes can be indicative of lowering the photosynthetic processes with relation to CuO-NPs toxicity (Rajput et al. 2015).

Plastoglobules are subcompartments of thylakoids that play an important role in lipid metabolic pathways (Austin et al. 2006), the chloroplast to chromoplast transition and the formation of colored carotenoid fibrils (Vishnevetsky et al. 1999). Previous studies showed an increased number of plastoglobules due to biotic, abiotic, and CuO-NPs-induced stress in *Landoltia punctate* (Lalau et al. 2015). The excess concentration of CuO-NPs severely affected starch content, stomatal aperture, epidermis, endodermis, cell wall, mitochondria, nuclei, and vascular bundles of *H. sativum* (Rajput et al. 2018a).

The identified changes in the root and leaf cell ultrastructure, especially in the photosynthetic apparatus, are associated with altered plant growth and performance.

6.3 Effects on Plant Physiology and Photosynthetic Systems

Photosynthesis is a key process for the conversion of light energy into chemical energy, which is performed by chloroplast, and other components of the photosynthetic machinery embedded in a highly dynamic matrix and thylakoid membranes (Rottet et al. 2015). Cu-based NPs may also affect photosynthesis and cause a decrease in electron transport, thylakoid number per granum, photosynthetic rate, transpiration rate, and stomatal conductance (Da Costa and Sharma 2015; Perreault et al. 2014). Musante and White (2012) observed that both bulk Cu and Cu-NPs reduced the transpiration rate by 60–70% in *C. pepo* relative to untreated controls. For the successful photochemical phenomena, chloroplast ultrastructure, thylakoid, grana formation, and physiological activities of photosynthetic machinery are important (Miller et al. 2017; Tighe-Neira et al. 2018). Thus, any structural and ultrastructural alteration in chloroplast apparatus and functionality associated with subcellular organelles such as plastoglobules and starch grains may adversely impact the overall photosynthesis (Fig. 2). Toxic effects of CuO-NPs were further shown in experiments with *O. sativa*. The CuO-NPs decreased Fv/Fm up to a complete loss of photosystem (PS) II photochemical quenching at a concentration of 1 mg L^{-1} and declined the photosynthetic pigment contents (Da Costa and Sharma 2015). It has further been reported that the CuO-NPs had a detrimental impact on the structure and function of the photosynthetic apparatus especially on photosynthetic pigments, chlorophyll, and grana (Tighe-Neira et al. 2018). Spring barley grown in hydroponic system showed accumulation of CuO-NPs in leaf cells and disorganized chloroplast structure and thylakoid in the mesophyll cells (Rajput et al. 2018a).

Thus, the declining photosynthetic efficiency can be a good forecaster of NPs toxic effects on plants.

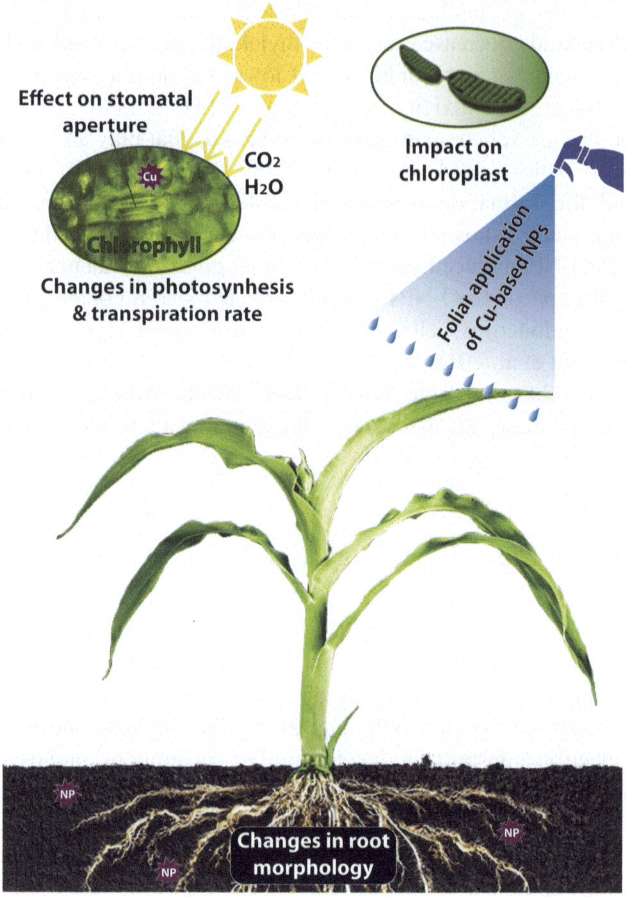

Fig. 2 Schematic presentation of effects of Cu-based NPs on photosynthesis

6.4 *Effects on Plant Metabolism and Nutrient Content*

Several studies have demonstrated that Cu-based NPs also significantly affect the metabolism and nutrient content of plants. For example, foliar application of Cu $(OH)_2$ nano-pesticide (50–1,000 nm) at 1,050–2,100 mg L^{-1} alters metabolite level of *L. sativa* leaves (Zhao et al. 2016b). Gas chromatography time-of-flight mass spectrometry (GCTOF-MS)-based analysis combined with partial least squares-discriminant analysis (PLS-DA) multivariate analysis shows disturbance in tricarboxylic acid (TCA) cycle and amino acid-related pathways (Zhao et al. 2016b, 2017b). An increased level of potassium, putrescine, and spermidine in $Cu(OH)_2$ nano-pesticide-treated plants has been suggested to reduce the oxidative stress and enhance the tolerance (Zhao et al. 2016b). Similarly, cucumber grown with Cu-NPs

(40 nm) in soil (200–800 mg kg^{-1}) and hydroponics (10 and 20 mg L^{-1}) exhibited perturbation in iron, sodium, phosphorus, zinc, sulfur, and molybdenum uptake and alterations in cucumber fruit metabolite profile (Zhao et al. 2016a). Additionally, TCA cycle and galactose metabolism also get compromised (Zhao et al. 2016b). CuO and $Cu(OH)_2$ nano-pesticides also decrease the level of shoot phosphorus and iron in lettuce (Hong et al. 2015). Moreover, CuO-NPs (<50 nm) at 500 mg kg^{-1} soil has shown to reduce iron, manganese, zinc, and calcium in common bean (Dimkpa et al. 2015). Moreover, micro- and macronutrients' elemental composition in cilantro has been found to be suppressed when grown with CuO-NPs (10^1–10^2 nm) and Cu-NPs (10^2–10^3 nm) at 0–80 mg kg^{-1} soil (Zuverza-Mena et al. 2015). The Cu-based NPs have also been documented to bring down the agronomically important characteristics of plants. The CuO-NPs (<50 nm) reduce carotenoid level in rice at 1 and 1.5 mM (Shaw and Hossain 2013). Similarly, the decrease in the firmness of cucumber fruits has been reported upon treatment with CuO-NPs of <50 nm (Hong et al. 2016). Also, the grain yield of rice has been shown to reduce significantly by CuO-NPs (~43 nm) at 500 and 1,000 mg kg^{-1} (Peng et al. 2017).

Summarizing these results, it can be concluded that Cu-based NPs at a certain concentration negatively affected plant metabolism and nutrient content.

6.5 Genotoxic and Cytotoxic Effects

Genotoxicity is one of the most devastating effects exerted by NPs on plants. A variety of toxic effects have been reported for NPs which may interact with biological systems via five main modes: (1) chemical effects as metal ions in solution upon dissolution; (2) mechanical effects owing to hard spheres and defined interfaces; (3) catalytic effects on surfaces; (4) surface effects owing to binding of proteins to the surface, either by non-covalent or covalent mechanisms or oxidative effects; and (5) changes in the chemical environment (pH). Metal and metal oxide NPs have been shown to act as mediators of DNA damage in mammalian cells, organisms, and even in bacteria, but the molecular mechanisms through which this occurs are poorly understood. For the first time, it was reported that CuO-NPs induce DNA damage in crops and grassland plants (Atha et al. 2012). The Cu-NPs, up to 20 μg mL^{-1}, increased the mitotic index of actively dividing cells in *A. cepa* with a gradual decline in the mitotic index as the concentration increased (Nagaonkar et al. 2015). Smaller-sized NPs, increasing concentrations, and exposure duration of NPs have been related to greater genotoxic responses, leading to mito-depressive effects in the cell cycle. Micronuclei formation, disturbed chromosomes, chromosome fragments, stickiness, bridge, laggards' chromosomes, and decrease in mitotic index are the most obvious anomalies in plants exposure to silver, copper, titanium dioxide, zinc, zinc oxide, selenium oxide, multi-wall carbon nanotube, tetramethylammonium hydroxide, and bismuth (III) oxide NPs. The severity of abnormalities depending on the concentration, duration time, and particle size is different. Finally, if the DNA repair mechanisms are not enough to restore these alterations, it can lead to loss of genetic material and mutation in DNA (Karami and de Lima 2016). The plant DNA is also affected by cellular oxidative stress generated by Cu-based NPs. Atha et al. (2012) reported oxidative stress-induced DNA lesions in *R. sativus*, *Lolium perenne*, and *L. rigidum* by CuO-NPs (10–1,000 mg L^{-1}) that include 2,6-diamino-4-hydroxy-5-formamidopyrimidine, 8-OH-dG, the 2′-deoxynucleoside form of 8-OH-G, and 4,6-diamino-5-formamidopyrimidine (Atha et al. 2012). Cu-based NPs exposure has been attributed to induce genotoxic effects and affect the normal cell cycle. Chromosomal aberrations such as sticky and disturbed chromosomes in metaphase/anaphase, c-metaphase, bridges, laggard chromosomes, disturbed telophase, and vacuolated nucleus resulted after exposure to Cu/CuO-NPs in onion and black cumin (Deng et al. 2016; Kumbhakar et al. 2016; Nagaonkar et al. 2015). These aberrations are very similar to those induced by ethyl methanesulfonate (EMS) and gamma radiation. With the use of random amplified polymorphic DNA (RAPD), the genotoxicity of CuO-NPs (~50 nm) has been demonstrated in buckwheat (Lee et al. 2013). The authors demonstrated changes in DNA bands in RAPD profiles of buckwheat exposed by 2,000 and 4,000 of CuO NPs mg L^{-1} (Lee et al. 2013). The changes in the genetic pattern induced by Cu-NPs toxicity could be attributed to changes in genomic DNA template stability due to mutations, homologous recombination, and deletion of large DNA segments and might be due to the strong binding of NPs with plant DNA (Ahmed et al. 2018b; Lee et al. 2013). The DNA isolated from young tomato leaves upon interaction with various concentrations of CuO-NPs exhibited concentration-dependent fluorescence quenching of acridine orange-DNA complex and ethidium bromide-DNA complex (Ahmed et al. 2018b). The CuO-NPs are able to interact with plant DNA in both intercalative and non-intercalative mode with perceptible changes in other macromolecules like amides I and II of proteins and carbohydrates (Ahmed et al. 2018b). The transfer of CuO-NPs to progeny (harvested seeds) of *Arabidopsis thaliana* has been studied by XANES in the form of CuO (88.8%); moreover, Cu in seeds has been detected as Cu-acetic acid (3.2%), $Cu_2(OH)PO_4$ (2%), and Cu_2O (6%) (Wang et al. 2016). Recently, the change in the gene expression pattern of plants exposed to CuO-NPs has been reported. Wang et al. (2016) documented differential expression of gene Fe-SOD and gene Aux/IAA in the regulation of *A. thaliana* root growth when exposed to 20 and 50 mg L^{-1} CuO-NPs. Similarly, altered gene expression has been observed by surface-enhanced laser desorption/ionization time-of-flight (SELDI-TOF) in cucumber seeds after treatment with nano-CuO at 600 mg L^{-1} (Moon et al. 2014). In this study, among 34 differentially expressed proteins about nine differed from those exposed to control and bulk CuO-treated plants. A protein (5,977-m/z) has been found as the most distinguished biomarker for the determination of CuO-NPs induced phytotoxicity (Moon et al. 2014).

Interaction of Cu-NPs with plant root exudates also influences the fate of Cu-NPs and magnitude of toxicity. Huang et al. (2017) determined the thermodynamic parameters for the interaction of Cu-NPs (40 nm) with a mixture of synthetic root exudates (SRE) and its components such as sugars, amino acids, organic acids, and phenolic acids by nano-isothermal titration calorimetry. The data revealed a strong binding constant ($K_d = 5.645 \times 10^3$ M^{-1}) for Cu-NPs SRE interaction; however, the binding of Cu^{2+} was found stronger but varied for individual SRE components (Huang et al. 2017).

The DNA damage and chromosomal aberrations raise the concern about the safety associated with applications of the NPs. However, the studies on the phytotoxicity of NPs are scarce, especially with regard to its mechanisms and on its potential uptake and subsequent fate within the food chain.

6.6 *Effects on Plant ROS and Anti-oxidative Activities*

One of the widely reported toxicity mechanisms is the generation of NP-induced ROS and consequent stimulation of cellular antioxidant defense mechanisms in plants. The NPs could enhance ROS generation in plants and cause oxidative stress, protein oxidation, lipid peroxidation, DNA damage, and finally cell death (Ahmed et al. 2018b; Mosa et al. 2018). To avoid oxidative stress, plants activate a defense mechanism involving the anti-oxidative enzymes (Rajput et al. 2015).

The ROS generation reportedly induces damage to cellular membranes resulting in respiratory loss and lipid peroxidation leading to disruption of vital cellular functions (Gueraud et al. 2010; Maness et al. 1999). In the presence of high concentrations, Cu can promote the generation of ROS by Fenton reaction ($Cu^+ + H_2O_2 \rightarrow Cu^{2+} + OH^\bullet + OH^-$) due to its high redox-active nature (Halliwell and Gutteridge 1985). ROS interaction with protein sulfhydryl (-SH) groups may cause enzyme inactivation which in all likelihood may lead to necrosis, chlorosis, and growth inhibition (Das and Roychoudhury 2014; Xiong and Wang 2005; Yruela 2009). Among ROS, hydroxyl radicals formed via Haber-Weiss reaction ($H_2O_2 + O_2^{\bullet-} \rightarrow OH^\bullet + OH^- + O_2$) are considered to be more toxic (Letelier et al. 2010). To mitigate the ROS stress induced by Cu-NPs, plants elevate the activity of antioxidant enzymes such as superoxide dismutase (SOD) (Wang et al. 2016), ascorbate peroxidase (APX) (Hong et al. 2015; Shaw et al. 2014), glutathione reductase (GR) (Shaw et al. 2014), catalase (CAT) (Ahmed et al. 2018a, b; Trujillo-Reyes et al. 2014), and peroxidase (POD) (Nair and Chung 2014). In addition to this, Cu-NPs arbitrated oxidative stress can also be measured in terms of antioxidant levels and proline (Shaw and Hossain 2013; Zhao et al. 2016b). The CuO-NPs exposure also increased the lipid peroxidation and triggered an imbalance in oxidative enzymes, viz., GSH, CAT, and POD (Dimkpa et al. 2012). The enhanced lipid peroxidation also accompanies low GSH and GSH/GSSG ratio (Shaw et al. 2014; Shaw and Hossain 2013) and high SOD activity that converts superoxide radicals into hydrogen peroxide ($O_2^\bullet \rightarrow H_2O_2$) (Kim et al. 2012; Nekrasova et al. 2011). Besides, antioxidant enzyme-enhanced malondialdehyde (MDA) content also serves as an oxidative stress marker for Cu-based NPs. For instance, the highest levels of MDA were observed in *C. sativus* shoots and roots treated with 100 and 200 and 50 and 100 mg L^{-1} Cu-NPs grown in a hydroponic system, respectively. An increase in MDA levels is directly proportional to the concentration of the Cu-NPs used for the treatment (Mosa et al. 2018). Similarly, the CuO-NPs increased lipid peroxidation and ROS in *Pisum sativum* (Nair and Chung 2015b).

To better understand the toxic nature of Cu-based NPs and their targeted applications, the end points of toxicity should be carefully scrutinized.

7 Toxicity on Aquatic Systems

The impact of Cu-based NPs on aquatic environment is an important issue due to extensive utilization of Cu-NPs, releasing metal ions in aqueous solution and making them bioavailable and toxic (Bondarenko et al. 2013; Chang et al. 2012; Mukherjee and Acharya 2018). The probabilistic model predicts environmental concentrations of Cu-NPs 0.06 mg L^{-1} in major Taiwanese rivers with 95% confidence interval (CI): 0.01–0.92 (Chio et al. 2012). This model raised concern on Cu-based NPs adverse effects on aquatic organisms. In addition, several studies highlighted toxicity of Cu-based NPs on aquatic organisms including gill injury and acute lethality in zebrafish and toxicity to algal species (Aruoja et al. 2009; Griffitt et al. 2007, 2009); induction of oxidative stress in the liver, gills, and muscles of juvenile *Epinephelus coioides* (Wang et al. 2014) and in mussels (Gomes et al. 2014); damage to gill filaments and gill pavement cells of freshwater fish (Song et al. 2015b); disruption of secondary lamellae of gills, damage in the liver showing pyknotic nuclei (Gupta et al. 2016); and affected proliferation, cell cycle progression, and cell death of amphibians (Thit et al. 2013). The summarized review on NPs toxicity on aquatic habitats suggests lethal effects on *Pseudokirchneriella subcapitata, Desmodesmus subspicatus, Xenopus laevis, Rana catesbeiana, Mytilus edulis, Mytilus galloprovincialis, Crassostrea virginica, Daphnia magna, Thamnocephalus platyurus, Danio rerio, Lytechinus pictus, Oncorhynchus mykiss*, and *Cyprinus carpio* (Mukherjee and Acharya 2018). Pradhan et al. (2015) found that CuO-NPs induce oxidative stress, damage to DNA, and plasma membrane of aquatic fungi. Similarly, Giannetto et al. (2018) found that CuO-NPs affected oxidative stress-related genes of *Arbacia lixula* embryos. A short-term study on diatom showed that Cu-NPs inhibited the growth, photosynthesis, and induced oxidative stress on *Phacodactylum tricornutum* (Zhu et al. 2017). Three different Lemnaceae species (*Spirodela polyrhiza, Lemna minor*, and *Wolffia arrhiza*) commonly found in freshwater lakes exposed to Cu-NPs expressed different sensitivities (Song et al. 2015a).

These data suggest that the toxicity of Cu-based NPs can be influenced by the species, exposure duration, and dose.

7.1 Toxicity on Aquatic Plants

There are potentially many sources of NPs in the aquatic ecosystem such as geogenic sources, industrial sources including medical and pharmaceutical, runoff from household's farms, leaching from landfills, etc. Xenobiotic substances could have a great impact on aquatic biota as well as constitute a serious danger for the aquatic ecosystem (Moore 2006). One of the anthropogenic sources of Cu-based NPs in the aquatic system is polymer coating found in marine paints or fabric with antimicrobial and biocidal properties. This kind of material is used for antifouling of boats and immersed structures, and CuO-NPs are frequently one of the ingredients (Almeida et al. 2007). A study showed that CuO-NPs alone (0.004 g L^{-1}) are less toxic to green alga *Chlamydomonas reinhardtii* than CuO-NPs coated with the polymer after 6 h of exposition (Melegari et al. 2013). Nonetheless, CuO-NPs still decreased the activity of PS II and were found responsible for the generation of ROS. There were observations for significantly higher intracellular Cu accumulation in the form of aggregate as compared to Cu-free samples (Perreault et al. 2012). Similar results were observed in the plant *Lemna gibba* such as morphological changes like abscission of the fronds from the colonies, decrease in frond size, and whitening of the fronds (Perreault et al. 2014). Both observations indicate that surface modification of NPs in order to enhance their stabilization changes their mechanism of toxicity which seems to be an important issue for expanding applications of Cu-based NPs in the future. Aruoja et al. (2009) performed tests on the bioavailability of Cu-based pollutants. The authors confirmed that Cu from CuO-NPs was 141-fold more bioavailable to aquatic flora in comparison to that from bulk CuO. The greater toxicity of CuO-NPs was seen in algae *Pseudokirchneriella* (Aruoja et al. 2009) and plant *Lemna minor* (Song et al. 2015a). That is consistent with the previous statement that the Cu bioavailability rather than the total concentration is the primary toxicity (Campbell 1995). However, Perreault et al. (2012) pose a hypothesis that during CuO-NPs solubilization, a soluble form of copper, mostly Cu^{2+} ions are released which can spread into the medium and become the main factor for CuO-NPs toxicity that is similar to the danger posed by $CuSO_4$. The *P. stratiotes* plants grown in the presence of Cu-NPs (1,000 mg L^{-1}) for 14 days exhibited discoloration along with the visible signs of turgor loss in mesophilic cells. Morphological changes in the root system were more prominent. In comparison to the control plant, blackening of roots together with inhibition of new growth roots and a decrease in plant weight, amino acids, and the content of ascorbic acid reduced by 63% were observed in exposed plants (Olkhovych et al. 2016). The morphological changes were also observed for plant *L. gibba* in the form of leaf reduction and detachment of fronds from the plant. The symptoms were detected after 24-h CuO-NPs exposure with 1.0 mg L^{-1} (Perreault et al. 2014). The growth inhibition was observed at 6.4 mg L^{-1} microalgae culture and for *L. minor* at 10 mg L^{-1} in comparison to Cu-free samples (Melegari et al. 2013; Song et al. 2016). The Cu-based NPs exposure on aquatic flora is mostly reflected in photosystem dysfunction. The chlorophyll content of *L. minor* decreased with the increase in concentration at 100 mg L^{-1} CuO-NPs (Song et al. 2016). In the algal culture of *C. reinhardti*, the decrease of total chlorophyll and carotenoids was observed at 1,000 mg L^{-1} when exposure lasted for 72 h (Aruoja et al. 2009). For microalgae, *Pseudokirchneriella* 6.4 mg L^{-1} was sufficient to evoke abnormality in photosynthetic system performance (Melegari et al. 2013). In the study of Perreault et al. (2014), lower photosynthetic electron transport rate for *L. giba* was observed. The

Cu-NPs at a concentration higher than 1 mg L^{-1} clearly suppress photosynthesis on *Elodea densa* (waterweed), while low concentration (<0.25 mg L^{-1}) has a positive impact on photosynthesis effectiveness (Nekrasova et al. 2011). The main feature of Cu-based NPs is that they have the ability to cross the plasma membrane that results in alteration of subcellular organelles. This condition substantially may cause oxidative stress which is connected to increased enzymatic activity (i.e., POD, CAT, and SOD) (Melegari et al. 2013). The production of ROS may be the result of conditions when plants are subjected to harmful stress conditions. The chloroplasts and mitochondria of plant cells are important in intracellular generators of ROS. Internal O_2 concentration is high during photosynthesis, and chloroplasts are particularly prone to generate ROS; therefore, these cytotoxic ROS can remarkably disrupt normal metabolism through oxidative damage of lipids, nucleic acids, and proteins.

In general Cu-based pollutants induce various responses within the photosynthetic organism. The changes seem to be the most prominent for the CuO-NPs and Cu-NPs following by $CuSO_4$ and bulk CuO. The Cu-NPs toxicity heavily depends on dosage and further surface modification.

7.2 Toxicity on Aquatic Animals

There is currently a significant gap in our knowledge about CuO-NPs toxicity to aquatic animals. In general, the Cu(O)-NPs toxicity may be a potential environmental concern for crustaceans, as LC50 values are within an order of magnitude of predicted wastewater concentrations, while chronic and developmental toxicity are a more relevant concern for fishes (Braz-Mota et al. 2018). A few studies have noted bioactivity in these animals at high concentrations (20 μg L^{-1}). The release of manufactured Cu-based NPs into the aquatic environment is rather rarely known (Moore 2006). Nevertheless, it was proven that NPs association with naturally occurring colloids may affect their bioavailability and uptake into cells and organisms. Uptake by endocytic routes was previously identified as probable major mechanisms of entry into cells, potentially leading to various types of toxic cell injury (Moore 2006). Griffitt et al. (2009) demonstrated that the effects of Cu-NPs were not solely due to the release of soluble metals into the water column. These studies highlight the need for further studies focused on understanding the mechanisms of NPs toxicity to aquatic organisms as dissolution, and the presence of a generic NPs response is not sufficient to explain the observed effects.

Sedimentation following hetero-aggregation with organic matter and free anions poses a threat due to benthic, sediment-dwelling and filter-feeding organisms. In marine systems, NPs can be absorbed by microorganisms and transferred to the next trophic levels by consumption. Filter feeders, especially bivalves, accumulate CuO-NPs through trapping them in mucus prior to ingestion. Benthic fauna may directly ingest sediment CuO-NPs. In fish, uptake is principally via the gut following drinking, while CuO-NPs caught in gill mucus may affect respiratory processes and ion transport. Currently, environmentally realistic CuO-NPs concentrations are unlikely to cause significant adverse acute health problems; however, sublethal effects, e.g., oxidative stress, have been noted in many organisms, often deriving from the dissolution of Cu^{2+}, and this could result in chronic health impacts (Baker et al. 2014).

The effect of waterborne Cu-NPs and copper sulfate on rainbow trout (*Oncorhynchus mykiss*) in the context of physiology and accumulation was also evaluated by Shaw et al. (2012). Overall, these data showed that Cu-NPs have similar types of toxic effects to $CuSO_4$, which can occur at lower tissue Cu concentrations than expected for the dissolved metal. It was also proved that CuO-NPs can induce toxicity to the freshwater shredder (*Allogamus ligonifer*) (Pradhan et al. 2012).

Abdel-Khalek et al. (2015) compared the toxicity of CuO-NPs to Nile tilapia (*Oreochromis niloticus*) with its bulk counterpart and reported that the LC50/96 h of CuO bulk particles (BPs) was higher than that of NPs indicating that CuO-NPs are more toxic. The CuO-NPs could exert more toxic effects despite the fact that they are smaller in size than the CuO-BPs, and they can form aggregates in suspensions. The authors demonstrated CuO (BPs and NPs)-induced biochemical alterations and oxidative stress in *O. niloticus*, which suggest ecological implications of CuO-NPs released in aquatic ecosystems. The study conducted by Braz-Mota et al. (2018) aimed to understand the effects of CuO-NPs and Cu on two ornamental Amazon fish species: dwarf cichlid (*Apistogramma agassizii*) and cardinal tetra (*Paracheirodon axelrodi*). For fish exposed to 50% of the LC50 for CuO-NPs, aerobic metabolic rate (MO_2), gill osmoregulatory physiology and mitochondrial function, oxidative stress markers, and morphological damage were evaluated. The results revealed species specificity in metabolic stress responses. An increase of MO_2 was noted in cardinal tetra exposed to Cu, but not CuO-NPs, whereas MO_2 in dwarf cichlid showed little change with either treatment. In contrast, mitochondria from dwarf cichlid exhibited increased proton leak and a resulting decrease in respiratory control ratios in response to CuO-NPs and Cu exposure. This uncoupling was directly related to an increase in ROS levels. The authors revealed different metabolic responses between these two species in response to CuO-NPs and Cu, which are probably caused by the differences between species' natural histories, indicating that different mechanisms of toxic action of the contaminants are associated to differential osmoregulatory strategies among species.

Gupta et al. (2016) described the effect of Cu-NPs exposure in the physiology of the common carp (*Cyprinus carpio*) using biochemical, histological, and proteomic approaches. The results indicated that the activity of oxidative stress enzymes catalase, superoxide dismutase, and glutathione-S-transferase was significantly increased in the kidney, liver, and gills of the treated groups when compared to control. Histological analysis revealed that after exposure, disruption of the secondary lamellae of gills, liver damage with pyknotic nuclei, and structural disarray of the kidney occurred. Proteomic analysis of the liver showed downregulation of several proteins including the ferritin heavy chain, Rho guanine nucleotide exchange factor 17-like, cytoglobin-1, regulation of diphosphomevalonate decarboxylase, and selenide and water dikinase-1.

The effect of Cu-NPs on the development of zebrafish embryos was depicted by Sun et al. (2016). The exposure to CuO-NPs at concentrations of 12.5 mg L^{-1} or higher leads to abnormal phenotypes and induces an inflammatory response in a dose-dependent pattern. Moreover, exposure to CuO-NPs at high doses results in an underdeveloped liver and a delay in retinal neurodifferentiation accompanied by reduced locomotor ability. The authors demonstrated that short-term exposure to CuO-NPs at high doses shows hepatotoxicity and neurotoxicity. On the other hand, cellular and molecular responses of adult zebrafish after exposure to CuO-NPs or ionic Cu were tested by Vicario-Pares et al. (2018). Another study performed by Bai et al. (2010) was undertaken to test the toxicity of nano-Cu suspension to zebrafish embryos. It was found that nano-Cu retarded the hatching of zebrafish embryos and caused morphological malformation of the larvae. The authors claimed that high concentrations (>0.1 mg L^{-1}) of nano-Cu can kill the gastrula-stage zebrafish embryos. Denluck et al. (2018) investigated the role of the chorion in nanomaterial toxicity. The authors found that the presence of the chorion inhibited Cu-NPs toxicity, while dechorionated embryonic zebrafish exposed to Cu-NPs had an LC50 of 2.5 ± 0.3 mg L^{-1} and a chorion-intact had LC50 of 13.7 ± 0.8 mg L^{-1}. In summary, embryo sensitivity increased by at least one order of magnitude when chorions were removed.

The toxicity of Cu-based NPs in aquatic environment appears to be one of the most important issues for assessing whole ecosystem safety. With no doubts, zebrafish embryos are excellent models for the study of nanomaterial-biological interactions and toxicity.

8 Techniques Used to Detect the Presence of Cu in Plant Tissues Treated with Cu-Based NPs

It has already been mentioned that new developments in nanotechnology industry increase the amount of such engineered nanomaterials in the environment, particularly in soils and aquatic ecosystems. This could lead to unpredicted consequences in the nearest future as plants play a vital role in the ecosystem and worldwide food supply. That is why NPs detection in environmental samples is of importance (Chaudhry et al. 2008; Mukherjee et al. 2016). However, not all methods are applicable to this problem due to low concentrations of NPs in environmental samples and experimental complications in sample preparation. Still, there are several modern techniques which are being widely applied to detect the presence, visualize the distribution, and analyze chemical properties of NPs in plant tissues or in the soil. The available detection methods could be classified into three broad sections: spectroscopy, diffraction, and imaging. However, the most comprehensive results could be obtained using the combination of all three methods. Besides, one of the most sensitive techniques is atomic absorption spectroscopy (AAS). However, this method is destructive and requires special sample preparation procedures.

Different types and combinations of electron microscopy techniques offer environmental scientists a wide range of capabilities. Scanning electron microscopy (SEM) gives a possibility to find and locate metal NPs which usually have higher electron density. SEM microscopes are often equipped with EDX that extend analytical capabilities to qualitative determination of elements present in the sample and quantitative determination of element concentration, thus opening a possibility to study the chemical composition of NPs. High-resolution TEM reveal the shape and morphology of tiny NPs of several nanometers in diameter. Selected area electron diffraction (SAED) and images acquired in bright and dark-field modes could be used to study NPs phase composition and distribution in the samples. Microscopes equipped with electron energy loss spectra (EELS) cameras are capable of revealing the oxidation state of 3D transition metals at nanoscale resolution (Tan et al. 2012). Moreover, these electron-based methods could be combined in one microscope that provides a great possibility to study the presence, distribution, chemical composition, morphology, shape, and size distribution of NPs in soils and plants. However, the shortcomings of the method are the limitations on the size of the sample, special sample preparation procedures, and the requirement of ultrahigh vacuum.

Furthermore, X-ray fluorescence (XRF) is one of the powerful tools to estimate the relative quantity of elements present in the sample semi-quantitatively (mass %). Often laboratory equipment has a focused X-ray beam up to 20–50 μm (μ-XRF) that gives a possibility to obtain element concentration maps of the samples with appropriate resolution. The latter could be used to detect and locate NPs aggregation in plants. There is also a particular interest in portable XRF devices (pXRF) (McLaren et al. 2012) for agronomic and environmental science applications as it opens possibilities to conduct in field studies. Such equipment could be used to relate plant conditions to elemental nutrient deficiencies in the soil (Towett et al. 2016). However, such devices are limited to spectroscopic data and low sensitivity. On the contrary, submicron resolution and high sensitivity of synchrotron-based micro- and nano-X-ray techniques open new possibilities to investigate the interactions between plants and engineered nanomaterials. Synchrotron-based techniques require minimal sample preparation, are nondestructive, and offer the best balance between sensitivity, chemical specificity, and spatial resolution (Castillo-Michel et al. 2017). These techniques are particularly adapted to investigate localization and speciation of NPs in plants: μ-XRF and synchrotron X-ray fluorescence mapping (SR-XFM) offers multi-elemental detection with resolution down to the tens of nm, in combination with spatially resolved X-ray absorption spectroscopy (μ-XAS or μ-XANES) speciation. Moreover, such synchrotron-based techniques could be combined with μ-XRD (micro X-ray diffraction) and μ-FTIR (micro Fourier-transform infrared spectroscopy) techniques in one beamline (Cotte et al. 2017).

One of the most promising methods to detect the presence of NPs at environmentally relevant concentration is sp-ICP-MS (Laborda et al. 2013, 2014). It gives a possibility to obtain qualitative information about the presence of particulate and/or dissolved forms, quantitative information as particle number as well as mass concentrations, and characterization information about the mass of element/s per particle and particle size (Laborda et al. 2016).

TEM remains one of the main tools to analyze Cu-based NPs distribution (Lee et al. 2008; Nhan Le et al. 2016) and composition in plants (Trujillo-Reyes et al. 2014; Wang et al. 2011). The XRF technique was applied to reveal the elemental composition of *C. sativus* shoot and root samples treated with Cu-NPs (Mosa et al. 2018). The microscopic analysis showed the presence of dense material in root cells of *O. sativum* L. treated with CuO-NPs and confirmed the presence of Cu by bulk-XANES, and the most dominant form of Cu was from CuO-NPs (Peng et al. 2015). A combination of μ-XRF and μ-XANES was used to study bioaccumulation un-weathered (U) and weathered (W) CuO-NPs, bulk and ionic form by lettuce (Servin et al. 2017b). The μ-XRF analysis of W-NP-exposed roots showed a homogenous distribution of Cu in the tissues, while μ-XANES analysis of W-NP-exposed roots showed near-complete transformation of CuO to Cu (I)-sulfur and oxide complexes in the tissues. Duran et al. (2017) showed that CuO-NPs did not affect seed germination of *Phaseolus vulgaris* L., but seedling weight gain was promoted by 100 mg Cu L^{-1} and inhibited by 1,000 mg Cu L^{-1} of 25 nm CuO and $CuSO_4$. The μ-XRF analysis showed that most of the Cu taken up remained in the seed coat with Cu hotspots in the hilum. Moreover, μ-XANES unraveled that most of Cu remained in its pristine form. Zhao et al. (2017a) showed significant growth inhibition on both roots and shoots of *E. crassipes* after 8-day exposure of CuO-NPs (50 mg L^{-1}) which was much higher than that of the bulk CuO particles and dissolved Cu^{2+} ions of the same Cu concentration. The XANES was used to reveal the presence of CuO-NPs as well as Cu_2S and other Cu species in roots, submerged leaves, and emerged leaves of plants providing solid evidence of the transformation of CuO-NPs. Electron microscopy remains one of the most widely used tools to study distribution, morphology, and composition of metal NPs in plants. The possibilities that such synchrotron radiation techniques as μ-XRF and μ-XANES open to environmental scientists could significantly change the situation in the sense of revealing precise information on its structure. Moreover, an sp-ICP-MS becomes one of the most promising technique to obtain the presence and size distribution of NPs at environmentally relevant concentrations.

9 Conclusion and Future Outlook

The literature unequivocally suggests that the higher concentrations of Cu-based NPs are detrimental to beneficial soil microorganisms, food crops, aquatic animals, and plants. The toxicity of Cu-based NPs is influenced by their composition, capping/coating material, size, and interactions with environmental components such as abiotic factors (e.g., pH) and microbial/plant secretions, naturally occurring organic matter, etc. Furthermore, the phytotoxicity may vary with the varying physiology/anatomy of plant species. Cu-based NPs are either taken up by organisms (internal efficiency) or adsorbed on external structures (external efficiency). The adherence and bioaccumulation may also be changed by physicochemical properties of Cu-based NPs, plant genotypes, and physical/chemical/biological transformation. The available studies considered in this review showed the inadequate characterization of Cu-based NPs, which could be the major obstacle in properly assessing its toxicity. Moreover, the disposal/discharge of Cu-based NPs into the environment is not regulated appropriately. After reviewing those studies, many questions still persist unanswered when the behavior and fate of Cu-based NPs in biological systems are taken into consideration. For instance, most of the studies on Cu-based NPs and plant interactions were performed on agar or in hydroponic media which do not reflect the actual interaction in the more realistic environment such as the soil system. The fate of Cu-based NPs and their toxicity and accumulation in the soil can vary significantly in different soil types due to the difference in pH, organic matter content, composition, etc. Therefore, understanding the connection between association and dissociation/dissolution of adequately characterized Cu-based NPs in a range of environmental media and the physiology/anatomy of affected organisms is most urgently needed to further our knowledge regarding the potential toxicity exerted by Cu-based NPs. After all, we conclude that Cu-based NPs comprised of Cu-NPs, CuO-NPs, and nano-Cu-based products used in agricultural practices have a great potential to negatively impact soil and aquatic micro-/macro-biota. The current scenario also emphasizes the regulated and safe dumping of waste containing Cu-based NPs into agroecosystems. In the future, the concentration of Cu-based NPs in edible parts of food crops must be measured carefully before supplying the products to consumers.

It is also crucial to develop a unified methodology for testing the NPs toxicity in natural environments. With the help of this methodology, joint research should be conducted to determine the toxicity of the same NPs under different climatic conditions and soil types. Such international research could help to develop the permissible levels of Cu-based NPs application and determine the threshold levels of their contents in different soils. The kinetics of NPs dissolution and migration to the groundwater should be specifically considered to avoid their accumulation above the safe levels. Sustainable use of Cu-based NPs could help to utilize the beneficial effects of their application (i.e., in the form of nanopesticides) without posing a threat to the living organisms.

The increased application of Cu-based NPs clearly indicates their negative impact on ecosystems. It is, therefore, imperative to explore Cu-based NPs toxicity and behavior in water, living organisms (biota), soil, and sediments individually and their toxicity in a combination of other metallic NPs. Past and future research must be placed in the context of current risk assessments associated with Cu-based NPs, their use, distribution, and release in the environment.

Acknowledgments This research was supported by the Ministry of Education and Science of Russia (project no. 5.948.2017/PP). We are grateful to Prof. Dr. Lok R. Pokhrel, Department of Public Health, East Carolina University, USA, for help in language editing.

Conflict of Interest The authors declare that they have no conflict of interest.

References

Abdel-Khalek AA, Kadry MAM, Badran SR, Marie MAS (2015) Comparative toxicity of copper oxide bulk and nano particles in Nile Tilapia; *Oreochromis niloticus*: biochemical and oxidative stress. J Basic Appl Zool 72:43–57

Adeleye AS, Conway JR, Perez T, Rutten P, Keller AA (2014) Influence of extracellular polymeric substances on the long-term fate, dissolution, and speciation of copper-based nanoparticles. Environ Sci Technol 48:12561–12568

Adeleye AS, Oranu EA, Tao M, Keller AA (2016) Release and detection of nanosized copper from a commercial antifouling paint. Water Res 102:374–382

Adhikari T, Dube G, Kundu S, Patra AK (2018) Impact of copper oxide nanoparticles on growth of different bacterial species. In: Energy and environment. Springer, Singapore, pp 47–55

Ahmed B, Dwivedi S, Abdin MZ, Azam A, Al-Shaeri M, Khan MS, Saquib Q, Al-Khedhairy AA, Musarrat J (2017) Mitochondrial and chromosomal damage induced by oxidative stress in $Zn^{(2+)}$ ions, ZnO-bulk and ZnO-NPs treated *Allium cepa* roots. Sci Rep 7:40685

Ahmed B, Hashmi A, Khan MS, Musarrat J (2018a) ROS mediated destruction of cell membrane, growth and biofilms of human bacterial pathogens by stable metallic AgNPs functionalized from bell pepper extract and quercetin. Adv Powder Technol 29:1601–1616

Ahmed B, Khan MS, Musarrat J (2018b) Toxicity assessment of metal oxide nano-pollutants on tomato (*Solanum lycopersicon*): a study on growth dynamics and plant cell death. Environ Pollut 240:802–816

Ahmed B, Khan MS, Saquib Q, Al-Shaeri M, Musarrat J (2018c) Interplay between engineered nanomaterials (ENMs) and edible plants: a current perspective. In: Faisal M, Saquib Q, Alatar AA, Al-Khedhairy AA (eds) Phytotoxicity of nanoparticles. Springer, Cham, pp 63–102

Ahmed B, Rizvi A, Zaidi A, Khan MS, Musarrat J (2019) Understanding the phyto-interaction of heavy metal oxide bulk and nanoparticles: evaluation of seed germination, growth, bioaccumulation, and metallothionein production. RSC Adv 9:4210–4225

Ali K, Ahmed B, Dwivedi S, Saquib Q, Al-Khedhairy AA, Musarrat J (2015) Microwave accelerated green synthesis of stable silver nanoparticles with *Eucalyptus globulus* leaf extract and their antibacterial and antibiofilm activity on clinical isolates. PLoS One 10:e0131178

Almeida E, Diamantino TC, de Sousa O (2007) Marine paints: the particular case of antifouling paints. Prog Org Coat 59:2–20

Anjum NA, Gill SS, Duarte AC, Pereira E, Ahmad I (2013) Silver nanoparticles in soil–plant systems. J Nanopart Res 15:1896

Anjum NA, Adam V, Kizek R, Duarte AC, Pereira E, Iqbal M, Lukatkin AS, Ahmad I (2015) Nanoscale copper in the soil-plant system – toxicity and underlying potential mechanisms. Environ Res 138:306–325

Anyaogu KC, Fedorov AV, Neckers DC (2008) Synthesis, characterization, and antifouling potential of functionalized copper nanoparticles. Langmuir 24:4340–4346

Arruda SC, Silva AL, Galazzi RM, Azevedo RA, Arruda MA (2015) Nanoparticles applied to plant science: a review. Talanta 131:693–705

Aruoja V, Dubourguier HC, Kasemets K, Kahru A (2009) Toxicity of nanoparticles of CuO, ZnO and TiO_2 to microalgae *Pseudokirchneriella subcapitata*. Sci Total Environ 407:1461–1148

Atha DH, Wang H, Petersen EJ, Cleveland D, Holbrook RD, Jaruga P, Dizdaroglu M, Xing B, Nelson BC (2012) Copper oxide nanoparticle mediated DNA damage in terrestrial plant models. Environ Sci Technol 46:1819–1827

Austin JR, Frost E, Vidi PA, Kessler F, Staehelin LA (2006) Plastoglobules are lipoprotein subcompartments of the chloroplast that are permanently coupled to thylakoid membranes and contain biosynthetic enzymes. Plant Cell 18:1693–1703

Bai W, Tian W, Zhang Z, He X, Ma Y, Liu N, Chai Z (2010) Effects of copper nanoparticles on the development of zebrafish embryos. J Nanosci Nanotechnol 10:8670–8676

Baker TJ, Tyler CR, Galloway TS (2014) Impacts of metal and metal oxide nanoparticles on marine organisms. Environ Pollut 186:257–271

Ben-Moshe T, Frenk S, Dror I, Minz D, Berkowitz B (2013) Effects of metal oxide nanoparticles on soil properties. Chemosphere 90:640–646

Ben-Sasson M, Lu X, Nejati S, Jaramillo H, Elimelech M (2016) In situ surface functionalization of reverse osmosis membranes with biocidal copper nanoparticles. Desalination 388:1–8

Bondarenko O, Juganson K, Ivask A, Kasemets K, Mortimer M, Kahru A (2013) Toxicity of Ag, CuO and ZnO nanoparticles to selected environmentally relevant test organisms and mammalian cells in vitro: a critical review. Arch Toxicol 87:1181–1200

Braz-Mota S, Campos DF, MacCormack TJ, Duarte RM, Val AL, Almeida-Val VMF (2018) Mechanisms of toxic action of copper and copper nanoparticles in two Amazon fish species: Dwarf cichlid (*Apistogramma agassizii*) and cardinal tetra (*Paracheirodon axelrodi*). Sci Total Environ 630:1168–1180

Buffet PE, Richard M, Caupos F, Vergnoux A, Perrein-Ettajani H, Luna-Acosta A, Akcha F, Amiard JC, Amiard-Triquet C, Guibbolini M, Risso-De Faverney C, Thomas-Guyon H, Reip P, Dybowska A, Berhanu D, Valsami-Jones E, Mouneyrac C (2013) A mesocosm study of fate and effects of CuO nanoparticles on endobenthic species (*Scrobicularia plana*, *Hediste diversicolor*). Environ Sci Technol 47:1620–1628

Bundschuh M, Filser J, Luderwald S, McKee MS, Metreveli G, Schaumann GE, Schulz R, Wagner S (2018) Nanoparticles in the environment: where do we come from, where do we go to? Environ Sci Eur 30:6

Campbell PGC (1995) Interactions between trace metals and aquatic organisms: a critique of the free-ion activity model. Wiley, New York

Castillo-Michel HA, Larue C, Pradas del Real AE, Cotte M, Sarret G (2017) Practical review on the use of synchrotron based micro- and nano- X-ray fluorescence mapping and X-ray absorption spectroscopy to investigate the interactions between plants and engineered nanomaterials. Plant Physiol Biochem 110:13–32

Chang YN, Zhang M, Xia L, Zhang J, Xing G (2012) The toxic effects and mechanisms of CuO and ZnO nanoparticles. Materials (Basel) 5:2850–2871

Chaudhry Q, Scotter M, Blackburn J, Ross B, Boxall A, Castle L, Aitken R, Watkins R (2008) Applications and implications of nanotechnologies for the food sector. Food Addit Contam 25:241–258

Chen Z, Gao SH, Jin M, Sun S, Lu J, Yang P et al (2019) Physiological and transcriptomic analyses reveal CuO nanoparticle inhibition of anabolic and catabolic activities of sulfate-reducing bacterium. Environ Int 125:65–74

Chio CP, Chen WY, Chou WC, Hsieh NH, Ling MP, Liao CM (2012) Assessing the potential risks to zebrafish posed by environmentally relevant copper and silver nanoparticles. Sci Total Environ 420:111–118

Colman BP, Arnaout CL, Anciaux S, Gunsch CK, Hochella MF Jr, Kim B, Lowry GV, McGill BM, Reinsch BC, Richardson CJ, Unrine JM, Wright JP, Yin L, Bernhardt ES (2013) Low concentrations of silver nanoparticles in biosolids cause adverse ecosystem responses under realistic field scenario. PLoS One 8:e57189

Concha-Guerrero SI, Brito EMS, Piñón-Castillo HA, Tarango-Rivero SH, Caretta CA, Luna-Velasco A, Duran R, Orrantia-Borunda E (2014) Effect of CuO nanoparticles over isolated

bacterial strains from agricultural soil. J Nanomater 2014:1–13. https://doi.org/10.1155/2014/148743
Conway JR, Beaulieu AL, Beaulieu NL, Mazer SJ, Keller AA (2015) Environmental stresses increase photosynthetic disruption by metal oxide nanomaterials in a soil-grown plant. ACS Nano 9:11737–11749
Cornelis G, Hund-Rinke K, Kuhlbusch T, van den Brink N, Nickel C (2014) Fate and bioavailability of engineered nanoparticles in soils: a review. Crit Rev Environ Sci Technol 44:2720–2764
Cota-Ruiz K, Delgado-Rios M, Martínez-Martínez A, Núñez-Gastelum JA, Peralta-Videa JR, Gardea-Torresdey JL (2018) Current findings on terrestrial plants-engineered nanomaterial interactions: are plants capable of phytoremediating nanomaterials from soil? Curr Opin Environ Sci Health 6:9–15
Cotte M et al (2017) The ID21 X-ray and infrared microscopy beamline at the ESRF: status and recent applications to artistic materials. J Anal At Spectrom 32:477–493
da Costa MVJ, Sharma PK (2015) Effect of copper oxide nanoparticles on growth, morphology, photosynthesis, and antioxidant response in *Oryza sativa*. Photosynthetica 54:110–119
Das K, Roychoudhury A (2014) Reactive oxygen species (ROS) and response of antioxidants as ROS-scavengers during environmental stress in plants. Front Environ Sci 2:53
del Real AEP, Castillo-Michel H, Kaegi R, Larue C, de Nolf W, Reyes-Herrera J et al (2018) Searching for relevant criteria to distinguish natural vs. anthropogenic TiO_2 nanoparticles in soils. Environ Sci Nano 5(12):2853–2863
Deng F, Wang S, Xin H (2016) Toxicity of CuO nanoparticles to structure and metabolic activity of *Allium cepa* root tips. Bull Environ Contam Toxicol 97:702–708
Denluck L, Wu F, Crandon LE, Harper BJ, Harper SL (2018) Reactive oxygen species generation is likely a driver of copper based nanomaterial toxicity. Environ Sci Nano 6:1473–1481
Dietz KJ, Herth S (2011) Plant nanotoxicology. Trends Plant Sci 16:582–589
Dimkpa CO, McLean JE, Latta DE, Manangón E, Britt DW, Johnson WP, Boyanov MI, Anderson AJ (2012) CuO and ZnO nanoparticles: phytotoxicity, metal speciation, and induction of oxidative stress in sand-grown wheat. J Nanopart Res 14:1125
Dimkpa CO, Latta DE, McLean JE, Britt DW, Boyanov MI, Anderson AJ (2013) Fate of CuO and ZnO nano- and microparticles in the plant environment. Environ Sci Technol 47:4734–4742
Dimkpa CO, McLean JE, Britt DW, Anderson AJ (2015) Nano-CuO and interaction with nano-ZnO or soil bacterium provide evidence for the interference of nanoparticles in metal nutrition of plants. Ecotoxicology 24:119–129
Du W, Tan W, Yin Y, Ji R, Peralta-Videa JR, Guo H, Gardea-Torresdey JL (2018) Differential effects of copper nanoparticles/microparticles in agronomic and physiological parameters of oregano (*Origanum vulgare*). Sci Total Environ 618:306–312
Duan X, Xu M, Zhou Y, Yan Z, Du Y, Zhang L, Zhang C, Bai L, Nie J, Chen G, Li F (2016) Effects of soil properties on copper toxicity to earthworm *Eisenia fetida* in 15 Chinese soils. Chemosphere 145:185–192
Duran NM, Savassa SM, Lima RG, de Almeida E, Linhares FS, van Gestel CAM, Pereira de Carvalho HW (2017) X-ray spectroscopy uncovering the effects of Cu based nanoparticle concentration and structure on *Phaseolus vulgaris* germination and seedling development. J Agric Food Chem 65:7874–7884
Feng Y, Cui X, He S, Dong G, Chen M, Wang J, Lin X (2013) The role of metal nanoparticles in influencing arbuscular mycorrhizal fungi effects on plant growth. Environ Sci Technol 47:9496–9504
Frenk S, Ben-Moshe T, Dror I, Berkowitz B, Minz D (2013) Effect of metal oxide nanoparticles on microbial community structure and function in two different soil types. PLoS One 8:e84441
Gao X, Avellan A, Laughton S, Vaidya R, Rodrigues SM, Casman EA, Lowry GV (2018) CuO nanoparticle dissolution and toxicity to wheat (*Triticum aestivum*) in Rhizosphere soil. Environ Sci Technol 52:2888–2897

Garner KL, Keller AA (2014) Emerging patterns for engineered nanomaterials in the environment: a review of fate and toxicity studies. J Nanopart Res 16:2503

Gautam A, Ray A, Mukherjee S, Das S, Pal K, Das S, Karmakar P, Ray M, Ray S (2018) Immunotoxicity of copper nanoparticle and copper sulfate in a common Indian earthworm. Ecotoxicol Environ Saf 148:620–631

Ge Y, Schimel JP, Holden PA (2011) Evidence for negative effects of TiO_2 and ZnO nanoparticles on soil bacterial communities. Environ Sci Technol 45:1659–1664

Giannetto A, Cappello T, Oliva S, Parrino V, De Marco G, Fasulo S, Mauceri A, Maisano M (2018) Copper oxide nanoparticles induce the transcriptional modulation of oxidative stress-related genes in *Arbacia lixula* embryos. Aquat Toxicol 201:187–197

Gogos A, Thalmann B, Voegelin A, Kaegi R (2017) Sulfidation kinetics of copper oxide nanoparticles. Environ Sci Nano 4:1733–1741

Gomes T, Chora S, Pereira CG, Cardoso C, Bebianno MJ (2014) Proteomic response of mussels *Mytilus galloprovincialis* exposed to CuO NPs and Cu(2)(+): an exploratory biomarker discovery. Aquat Toxicol 155:327–336

Griffitt RJ, Weil R, Hyndman KA, Denslow ND, Powers K, Taylor D, Barber DS (2007) Exposure to copper nanoparticles causes gill injury and acute lethality in zebrafish (*Danio rerio*). Environ Sci Technol 41:8178–8186

Griffitt RJ, Luo J, Gao J, Bonzongo JC, Barber DS (2009) Effects of particle composition and species on toxicity of metallic nanomaterials in aquatic organisms. Environ Toxicol Chem 27:1972–1978

Gueraud F, Atalay M, Bresgen N, Cipak A, Eckl PM, Huc L, Jouanin I, Siems W, Uchida K (2010) Chemistry and biochemistry of lipid peroxidation products. Free Radic Res 44:1098–1124

Gupta YR, Sellegounder D, Kannan M, Deepa S, Senthilkumaran B, Basavaraju Y (2016) Effect of copper nanoparticles exposure in the physiology of the common carp (*Cyprinus carpio*): biochemical, histological and proteomic approaches. Aquac Fish 1:15–23

Halliwell B, Gutteridge JMC (1985) Free radicals in biology and medicine. Free Radic Biol Med 1:331–332

Hanna SK, Miller RJ, Zhou D, Keller AA, Lenihan HS (2013) Accumulation and toxicity of metal oxide nanoparticles in a soft-sediment estuarine amphipod. Aquat Toxicol 142–143:441–446

Haris Z, Ahmad I (2017) Impact of metal oxide nanoparticles on beneficial soil microorganisms and their secondary metabolites. Int J Sci Res 3:1020–1030

He S, Feng Y, Ni J, Sun Y, Xue L, Feng Y, Yu Y, Lin X, Yang L (2016) Different responses of soil microbial metabolic activity to silver and iron oxide nanoparticles. Chemosphere 147:195–202

Hong J, Rico CM, Zhao L, Adeleye AS, Keller AA, Peralta-Videa JR, Gardea-Torresdey JL (2015) Toxic effects of copper-based nanoparticles or compounds to lettuce (*Lactuca sativa*) and alfalfa (*Medicago sativa*). Environ Sci Process Impacts 17:177–185

Hong J, Wang L, Sun Y, Zhao L, Niu G, Tan W, Rico CM, Peralta-Videa JR, Gardea-Torresdey JL (2016) Foliar applied nanoscale and microscale CeO_2 and CuO alter cucumber (*Cucumis sativus*) fruit quality. Sci Total Environ 563–564:904–911

Huang Y, Zhao L, Keller AA (2017) Interactions, transformations, and bioavailability of nano-copper exposed to root exudates. Environ Sci Technol 51:9774–9783

Ingle AP, Duran N, Rai M (2014) Bioactivity, mechanism of action, and cytotoxicity of copper-based nanoparticles: a review. Appl Microbiol Biotechnol 98:1001–1009

Karami MS, de Lima R (2016) Nanoparticles cyto and genotoxicity in plants: mechanisms and abnormalities. Environ Nanotechnol Monit Manage 6:184–193

Kasemets K, Ivask A, Dubourguier HC, Kahru A (2009) Toxicity of nanoparticles of ZnO, CuO and TiO_2 to yeast *Saccharomyces cerevisiae*. Toxicol In Vitro 23:1116–1122

Katsumiti A, Thorley AJ, Arostegui I, Reip P, Valsami-Jones E, Tetley TD, Cajaraville MP (2018) Cytotoxicity and cellular mechanisms of toxicity of CuO NPs in mussel cells in vitro and comparative sensitivity with human cells. Toxicol In Vitro 48:146–158

Keller AA, Lazareva A (2013) Predicted releases of engineered nanomaterials: from global to regional to local. Environ Sci Technol Lett 1:65–70

Keller AA, McFerran S, Lazareva A, Suh S (2013) Global life cycle releases of engineered nanomaterials. J Nanopart Res 15:1692

Keller AA, Adeleye AS, Conway JR, Garner KL, Zhao L, Cherr GN, Hong J, Gardea-Torresdey JL, Godwin HA, Hanna S, Ji Z, Kaweeteerawat C, Lin S, Lenihan HS, Miller RJ, Nel AE, Peralta-Videa JR, Walker SL, Taylor AA, Torres-Duarte C, Zink JI, Zuverza-Mena N (2017) Comparative environmental fate and toxicity of copper nanomaterials. NanoImpact 7:28–40

Keller AA, Huang Y, Nelson J (2018) Detection of nanoparticles in edible plant tissues exposed to nano copper using single particle ICP-MS. J Nanopart Res 20:101

Kim S, Lee S, Lee I (2012) Alteration of phytotoxicity and oxidant stress potential by metal oxide nanoparticles in *Cucumis sativus*. Water Air Soil Pollut 223:2799–2806

Kim S, Sin H, Lee S, Lee I (2013) Influence of metal oxide particles on soil enzyme activity and bioaccumulation of two plants. J Microbiol Biotechnol 23:1279–1286

Kovacec E, Regvar M, van Elteren JT, Arcon I, Papp T, Makovec D, Vogel-Mikus K (2017) Biotransformation of copper oxide nanoparticles by the pathogenic fungus *Botrytis cinerea*. Chemosphere 180:178–185

Kumar N, Shah V, Walker VK (2012) Influence of a nanoparticle mixture on an arctic soil community. Environ Toxicol Chem 31:131–135

Kumbhakar DV, Datta AK, Mandal A, Das D, Gupta S, Ghosh B, Halder S, Dey S (2016) Effectivity of copper and cadmium sulphide nanoparticles in mitotic and meiotic cells of *Nigella sativa* L. (black cumin) – can nanoparticles act as mutagenic agents? J Exp Nanosci 11:823–839

Laborda F, Jimenez-Lamana J, Bolea E, Castillo JR (2013) Critical considerations for the determination of nanoparticle number concentrations, size and number size distributions by single particle ICP-MS. J Anal At Spectrom 28:1220–1232

Laborda F, Bolea E, Jimenez-Lamana J (2014) Single particle inductively coupled plasma mass spectrometry: a powerful tool for nanoanalysis. Anal Chem 86:2270–2278

Laborda F, Bolea E, Jimenez-Lamana J (2016) Single particle inductively coupled plasma mass spectrometry for the analysis of inorganic engineered nanoparticles in environmental samples. Trends Environ Anal Chem 9:15–23

Lalau CM, Mohedano Rde A, Schmidt EC, Bouzon ZL, Ouriques LC, dos Santos RW, da Costa CH, Vicentini DS, Matias WG (2015) Toxicological effects of copper oxide nanoparticles on the growth rate, photosynthetic pigment content, and cell morphology of the duckweed Landoltia punctata. Protoplasma 252:221–229

Lee WM, An YJ, Yoon H, Kweon HS (2008) Toxicity and bioavailability of copper nanoparticles to the terrestrial plants mung bean (*Phaseolus radiatus*) and wheat (*Triticum aestivum*): plant agar test for water-insoluble nanoparticles. Environ Toxicol Chem 27:1915–1921

Lee S, Chung H, Kim S, Lee I (2013) The genotoxic effect of ZnO and CuO nanoparticles on early growth of buckwheat, *Fagopyrum esculentum*. Water Air Soil Pollut 224:1668

Letelier ME, Sanchez-Jofre S, Peredo-Silva L, Cortes-Troncoso J, Aracena-Parks P (2010) Mechanisms underlying iron and copper ions toxicity in biological systems: pro-oxidant activity and protein-binding effects. Chem Biol Interact 188:220–227

Liu Y, Baas J, Peijnenburg WJ, Vijver MG (2016) Evaluating the combined toxicity of Cu and ZnO nanoparticles: utility of the concept of additivity and a nested experimental design. Environ Sci Technol 50:5328–5337

Lofts S, Criel P, Janssen CR, Lock K, McGrath SP, Oorts K, Rooney CP, Smolders E, Spurgeon DJ, Svendsen C, van Eeckhout H, Zhao FZ (2013) Modelling the effects of copper on soil organisms and processes using the free ion approach: towards a multi-species toxicity model. Environ Pollut 178:244–253

Lowry GV, Gregory KB, Apte SC, Lead JR (2012) Transformations of nanomaterials in the environment. Environ Sci Technol 46:6893–6899

Ma X, Geisler-Lee J, Deng Y, Kolmakov A (2010) Interactions between engineered nanoparticles (ENPs) and plants: phytotoxicity, uptake and accumulation. Sci Total Environ 408:3053–3061

Ma R, Stegemeier J, Levard C, Dale JG, Noack CW, Yang T, Brown GE, Lowry GV (2014) Sulfidation of copper oxide nanoparticles and properties of resulting copper sulfide. Environ Sci Nano 1:347–357

Manceau A, Nagy KL, Marcus MA, Lanson M, Geoffroy N, Jacquet T, Kirpichtchikova T (2008) Formation of metallic copper nanoparticles at the soil-root interface. Environ Sci Technol 42:1766–1772

Maness PC, Smolinski S, Blake DM, Huang Z, Wolfrum EJ, Jacoby WA (1999) Bactericidal activity of photocatalytic TiO(2) reaction: toward an understanding of its killing mechanism. Appl Environ Microbiol 65:4094–4098

Manusadzianas L, Caillet C, Fachetti L, Gylyte B, Grigutyte R, Jurkoniene S, Karitonas R, Sadauskas K, Thomas F, Vitkus R, Ferard JF (2012) Toxicity of copper oxide nanoparticle suspensions to aquatic biota. Environ Toxicol Chem 31:108–114

Mashock MJ, Zanon T, Kappell AD, Petrella LN, Andersen EC, Hristova KR (2016) Copper oxide nanoparticles impact several toxicological endpoints and cause neurodegeneration in *Caenorhabditis elegans*. PLoS One 11:e0167613

McLaren TI, Guppy CN, Tighe MK (2012) A rapid and nondestructive plant nutrient analysis using portable X-ray fluorescence. Soil Sci Soc Am J 76:1446–1453

Melegari SP, Perreault F, Costa RH, Popovic R, Matias WG (2013) Evaluation of toxicity and oxidative stress induced by copper oxide nanoparticles in the green alga *Chlamydomonas reinhardtii*. Aquat Toxicol 142–143:431–440

Miao L, Wang C, Hou J, Wang P, Ao Y, Li Y, Lv B, Yang Y, You G, Xu Y (2015) Enhanced stability and dissolution of CuO nanoparticles by extracellular polymeric substances in aqueous environment. J Nanopart Res 17:404

Miller RJ, Muller EB, Cole B, Martin T, Nisbet R, Bielmyer-Fraser GK, Jarvis TA, Keller AA, Cherr G, Lenihan HS (2017) Photosynthetic efficiency predicts toxic effects of metal nanomaterials in phytoplankton. Aquat Toxicol 183:85–93

Montes M, Pierce CG, Lopez-Ribot JL, Bhalla AS, Guo RY (2016) Properties of silver and copper nanoparticle containing aqueous suspensions and evaluation of their in vitro activity against *Candida albicans* and *Staphylococcus aureus* biofilms. J Nanopart Res 37:109–121

Moon YS, Park ES, Kim TO, Lee HS, Lee SE (2014) SELDI-TOF MS-based discovery of a biomarker in *Cucumis sativus* seeds exposed to CuO nanoparticles. Environ Toxicol Pharmacol 38:922–931

Moore MN (2006) Do nanoparticles present ecotoxicological risks for the health of the aquatic environment? Environ Int 32:967–976

Mosa KA, El-Naggar M, Ramamoorthy K, Alawadhi H, Elnaggar A, Wartanian S, Ibrahim E, Hani H (2018) Copper nanoparticles induced genotoxicity, oxidative stress, and changes in superoxide dismutase (SOD) gene expression in cucumber (*Cucumis sativus*) plants. Front Plant Sci 9:872

Mudunkotuwa IA, Pettibone JM, Grassian VH (2012) Environmental implications of nanoparticle aging in the processing and fate of copper-based nanomaterials. Environ Sci Technol 46:7001–7010

Mukherjee K, Acharya K (2018) Toxicological effect of metal oxide nanoparticles on soil and aquatic habitats. Arch Environ Contam Toxicol 75:175–186

Mukherjee A, Peralta-Videa JR, Gardea-Torresdey J (2016) Effects and uptake of nanoparticles in plants. In: Baoshan Xing CDV, Senesi N (eds) Engineered nanoparticles and the environment: biophysicochemical processes and toxicity. Wiley, Hoboken, pp 386–408

Musante C, White JC (2012) Toxicity of silver and copper to Cucurbita pepo: differential effects of nano and bulk-size particles. Environ Toxicol 27:510–517

Nagajyoti PC, Lee KD, Sreekanth TVM (2010) Heavy metals, occurrence and toxicity for plants: a review. Environ Chem Lett 8:199–216

Nagaonkar D, Shende S, Rai M (2015) Biosynthesis of copper nanoparticles and its effect on actively dividing cells of mitosis in *Allium cepa*. Biotechnol Prog 31:557–565

Nair PM, Chung IM (2014) A mechanistic study on the toxic effect of copper oxide nanoparticles in soybean (*Glycine max* L.) root development and lignification of root cells. Biol Trace Elem Res 162:342–352

Nair PM, Chung IM (2015a) Study on the correlation between copper oxide nanoparticles induced growth suppression and enhanced lignification in Indian mustard (*Brassica juncea* L.). Ecotoxicol Environ Saf 113:302–313

Nair PMG, Chung IM (2015b) The responses of germinating seedlings of green peas to copper oxide nanoparticles. Biol Plant 59:591–595

Nekrasova GF, Ushakova OS, Ermakov AE, Uimin MA, Byzov IV (2011) Effects of copper (II) ions and copper oxide nanoparticles on *Elodea densa* planch. Russ J Ecol 42:458–463

Nhan Le V, Ma C, Shang J, Rui Y, Liu S, Xing B (2016) Effects of CuO nanoparticles on insecticidal activity and phytotoxicity in conventional and transgenic cotton. Chemosphere 144:661–670

Ojha NK, Zyryanov GV, Majee A, Charushin VN, Chupakhin ON, Santra S (2017) Copper nanoparticles as inexpensive and efficient catalyst: a valuable contribution in organic synthesis. Corros Rev 353:1–57

Olkhovych O, Volkogon M, Taran N, Batsmanova L, Kravchenko I (2016) The effect of copper and zinc nanoparticles on the growth parameters, contents of ascorbic acid, and qualitative composition of amino acids and acylcarnitines in *Pistia stratiotes* L. (Araceae). Nanoscale Res Lett 11:218

Parada J, Rubilar O, Diez MC, Cea M, da Silva ASA, Rodríguez-Rodríguez CE, Tortella GR (2019) Combined pollution of copper nanoparticles and atrazine in soil: effects on dissipation of the pesticide and on microbiological community profiles. J Hazard Mater 361:228–236

Parra B, Tortella GR, Cuozzo S, Martínez M (2019) Negative effect of copper nanoparticles on the conjugation frequency of conjugative catabolic plasmids. Ecotoxicol Environ Saf 169:662–668

Peng C, Duan D, Xu C, Chen Y, Sun L, Zhang H, Yuan X, Zheng L, Yang Y, Yang J, Zhen X, Chen Y, Shi J (2015) Translocation and biotransformation of CuO nanoparticles in rice (*Oryza sativa* L.) plants. Environ Pollut 197:99–107

Peng C, Xu C, Liu Q, Sun L, Luo Y, Shi J (2017) Fate and transformation of CuO nanoparticles in the soil-rice system during the life cycle of rice plants. Environ Sci Technol 51:4907–4917

Perreault F, Oukarroum A, Melegari SP, Matias WG, Popovic R (2012) Polymer coating of copper oxide nanoparticles increases nanoparticles uptake and toxicity in the green alga *Chlamydomonas reinhardtii*. Chemosphere 87:1388–1394

Perreault F, Popovic R, Dewez D (2014) Different toxicity mechanisms between bare and polymer-coated copper oxide nanoparticles in *Lemna gibba*. Environ Pollut 185:219–227

Philippe A, Schaumann GE (2014) Interactions of dissolved organic matter with natural and engineered inorganic colloids: a review. Environ Sci Technol 48:8946–8962

Ponmurugan P, Manjukarunambika K, Elango V, Gnanamangai BM (2016) Antifungal activity of biosynthesised copper nanoparticles evaluated against red root-rot disease in tea plants. J Exp Nanosci 11:1019–1031

Pradhan A, Seena S, Pascoal C, Cassio F (2011) Can metal nanoparticles be a threat to microbial decomposers of plant litter in streams? Microb Ecol 62:58–68

Pradhan A, Seena S, Pascoal C, Cassio F (2012) Copper oxide nanoparticles can induce toxicity to the freshwater shredder *Allogamus ligonifer*. Chemosphere 89:1142–1150

Pradhan A, Seena S, Schlosser D, Gerth K, Helm S, Dobritzsch M, Krauss GJ, Dobritzsch D, Pascoal C, Cassio F (2015) Fungi from metal-polluted streams may have high ability to cope with the oxidative stress induced by copper oxide nanoparticles. Environ Toxicol Chem 34:923–930

Pullagurala VLR, Rawat S, Adisa IO, Hernandez-Viezcas JA, Peralta-Videa JR, Gardea-Torresdey JL (2018) Plant uptake and translocation of contaminants of emerging concern in soil. Sci Total Environ 636:1585–1596

Qiu H, Smolders E (2017) Nanospecific phytotoxicity of CuO nanoparticles in soils disappeared when bioavailability factors were considered. Environ Sci Technol 51:11976–11985

Rajput VD, Chen Y, Ayup M (2015) Effects of high salinity on physiological and anatomical indices in the early stages of Populus euphratica growth. Russ J Plant Physiol 62:229–236

Rajput VD, Minkina T, Sushkova S, Tsitsuashvili V, Mandzhieva S, Gorovtsov A, Nevidomskyaya D, Gromakova N (2017a) Effect of nanoparticles on crops and soil microbial communities. J Soil Sediment 18:2179–2187

Rajput VD, Minkina T, Suskova S, Mandzhieva S, Tsitsuashvili V, Chapligin V, Fedorenko A (2017b) Effects of copper nanoparticles (CuO NPs) on crop plants: a mini review. BioNanoSci 8:36–42

Rajput V, Minkina T, Fedorenko A, Sushkova S, Mandzhieva S, Lysenko V, Duplii N, Fedorenko G, Dvadnenko K, Ghazaryan K (2018a) Toxicity of copper oxide nanoparticles on spring barley (*Hordeum sativum* distichum). Sci Total Environ 645:1103–1113

Rajput VD, Minkina T, Fedorenko A, Tsitsuashvili V, Mandzhieva S, Sushkova S, Azarov A (2018b) Metal oxide nanoparticles: applications and effects on soil ecosystems. In: Soil contamination: sources, assessment and remediation. Nova Science Publishers, Hauppauge, pp 81–106

Rajput VD, Minkina TM, Behal A, Sushkova SN, Mandzhieva S, Singh R, Gorovtsov A, Tsitsuashvili VS, Purvis WO, Ghazaryan KA, Movsesyan HS (2018c) Effects of zinc-oxide nanoparticles on soil, plants, animals and soil organisms: a review. Environ Nanotechnol Monit 9:76–84

Rajput VD, Minkina T, Fedorenko A, Mandzhieva S, Sushkova S, Lysenko V, Duplii N, Azarov A, Chokheli V (2018d) Destructive effect of copper oxide nanoparticles on ultrastructure of chloroplast, plastoglobules and starch grains in spring barley (*Hordeum sativum* distichum). Int J Agric Biol 21:171–174. https://doi.org/10.17957/IJAB/15.0877

Raliya R, Franke C, Chavalmane S, Nair R, Reed N, Biswas P (2016) Quantitative understanding of nanoparticle uptake in watermelon plants. Front Plant Sci 7:1288

Rastogi A, Zivcak M, Sytar O, Kalaji HM, He X, Mbarki S, Brestic M (2017) Impact of metal and metal oxide nanoparticles on plant: a critical review. Front Chem 5:78

Ray D, Pramanik S, Prasad MR, Chaudhuri S, De S (2015) Sugar-mediated 'green' synthesis of copper nanoparticles with high antifungal activity. Mater Res Express 2:105002

Rico CM, Majumdar S, Duarte-Gardea M, Peralta-Videa JR, Gardea-Torresdey JL (2011) Interaction of nanoparticles with edible plants and their possible implications in the food chain. J Agric Food Chem 59:3485–3498

Rico CM, Peralta-Videa JR, Gardea-Torresdey JL (2015) Differential effects of cerium oxide nanoparticles on rice, wheat, and barley roots: a fourier transform infrared (FT-IR) microspectroscopy study. Appl Spectrosc 69:287–295

Rottet S, Besagni C, Kessler F (2015) The role of plastoglobules in thylakoid lipid remodeling during plant development. Biochim Biophys Acta 1847:889–899

Rui M, Ma C, White JC, Hao Y, Wang Y, Tang X, Yang J, Jiang F, Ali A, Rui Y, Cao W, Chen G, Xing B (2018) Metal oxide nanoparticles alter peanut (*Arachis hypogaea* L.) physiological response and reduce nutritional quality: a life cycle study. Environ Sci-Nano 5:2088–2102

Ruiz P, Katsumiti A, Nieto JA, Bori J, Jimeno-Romero A, Reip P, Arostegui I, Orbea A, Cajaraville MP (2015) Short-term effects on antioxidant enzymes and long-term genotoxic and carcinogenic potential of CuO nanoparticles compared to bulk CuO and ionic copper in mussels *Mytilus galloprovincialis*. Mar Environ Res 111:107–120

Saleem S, Ahmed B, Khan MS, Al-Shaeri M, Musarrat J (2017) Inhibition of growth and biofilm formation of clinical bacterial isolates by NiO nanoparticles synthesized from *Eucalyptus globulus* plants. Microb Pathog 111:375–387

Sedighi A, Montazer M (2016) Tunable shaped N-doped CuO nanoparticles on cotton fabric through processing conditions: synthesis, antibacterial behavior and mechanical properties. Cellul 23:2229–2243

Servin AD, De la Torre-Roche R, Castillo-Michel H, Pagano L, Hawthorne J, Musante C, Pignatello J, Uchimiya M, White JC (2017a) Exposure of agricultural crops to nanoparticle CeO_2 in biochar-amended soil. Plant Physiol Biochem 110:147–157

Servin AD, Pagano L, Castillo-Michel H, De la Torre-Roche R, Hawthorne J, Hernandez-Viezcas JA, Loredo-Portales R, Majumdar S, Gardea-Torresday J, Dhankher OP, White JC (2017b) Weathering in soil increases nanoparticle CuO bioaccumulation within a terrestrial food chain. Nanotoxicology 11:98–111

Shah V, Luxton TP, Walker VK, Brumfield T, Yost J, Shah S, Wilkinson JE, Kambhampati M (2016) Fate and impact of zero-valent copper nanoparticles on geographically-distinct soils. Sci Total Environ 573:661–670

Shahid M, Khan MS (2017) Assessment of glyphosate and quizalofop mediated toxicity to greengram (*Vigna radiata* (L.) Wilczek), stress abatement and growth promotion by herbicide tolerant bradyrhizobium and pseudomonas species. Int J Curr Microbiol App Sci 6:3001–3016

Shaw AK, Hossain Z (2013) Impact of nano-CuO stress on rice (*Oryza sativa* L.) seedlings. Chemosphere 93:906–915

Shaw BJ, Al-Bairuty G, Handy RD (2012) Effects of waterborne copper nanoparticles and copper sulphate on rainbow trout, (*Oncorhynchus mykiss*): physiology and accumulation. Aquat Toxicol 116–117:90–101

Shaw AK, Ghosh S, Kalaji HM, Bosa K, Brestic M, Zivcak M, Hossain Z (2014) Nano-CuO stress induced modulation of antioxidative defense and photosynthetic performance of Syrian barley (*Hordeum vulgare* L.). Environ Exp Bot 102:37–47

Shi J, Ye J, Fang H, Zhang S, Xu C (2018) Effects of copper oxide nanoparticles on paddy soil properties and components. Nano 8(10):839

Simonin M, Richaume A (2015) Impact of engineered nanoparticles on the activity, abundance, and diversity of soil microbial communities: a review. Environ Sci Pollut Res Int 22:13710–13723

Simonin M, Cantarel AA, Crouzet A, Gervaix J, Martins JM, Richaume A (2018) Negative effects of copper oxide nanoparticles on carbon and nitrogen cycle microbial activities in contrasting agricultural soils and in presence of plants. Front Microbiol 9:3102

Slotte M, Zevenhoven R (2017) Energy requirements and life cycle assessment of production and product integration of silver, copper and zinc nanoparticles. J Clean Prod 148:948–957

Song L, Vijver MG, Peijnenburg WJ (2015a) Comparative toxicity of copper nanoparticles across three Lemnaceae species. Sci Total Environ 518–519:217–224

Song L, Vijver MG, Peijnenburg WJ, Galloway TS, Tyler CR (2015b) A comparative analysis on the in vivo toxicity of copper nanoparticles in three species of freshwater fish. Chemosphere 139:181–189

Song G, Hou W, Gao Y, Wang Y, Lin L, Zhang Z, Niu Q, Ma R, Mu L, Wang H (2016) Effects of CuO nanoparticles on *Lemna minor*. Bot Stud 57:3

Stampoulis D, Sinha SK, White JC (2009) Assay-dependent phytotoxicity of nanoparticles to plants. Environ Sci Technol 43:9473–9479

Sun Y, Zhang G, He Z, Wang Y, Cui J, Li Y (2016) Effects of copper oxide nanoparticles on developing zebrafish embryos and larvae. Int J Nanomedicine 11:905–918

Sun L, Yang J, Fang H, Xu C, Peng C, Huang H, Lu L, Duan D, Zhang X, Shi J (2017) Mechanism study of sulfur fertilization mediating copper translocation and biotransformation in rice (*Oryza sativa* L.) plants. Environ Pollut 226:426–434

Tan HY, Verbeeck J, Abakumov A, van Tendeloo G (2012) Oxidation state and chemical shift investigation in transition metal oxides by EELS. Ultramicroscopy 116:24–33

Thit A, Selck H, Bjerregaard HF (2013) Toxicity of CuO nanoparticles and Cu ions to tight epithelial cells from *Xenopus laevis* (A6): effects on proliferation, cell cycle progression and cell death. Toxicol In Vitro 27:1596–1601

Tighe-Neira R, Carmora E, Recio G, Nunes-Nesi A, Reyes-Diaz M, Alberdi M, Rengel Z, Inostroza-Blancheteau C (2018) Metallic nanoparticles influence the structure and function of the photosynthetic apparatus in plants. Plant Physiol Biochem 130:408–417

Torres-Duarte C, Adeleye AS, Pokhrel S, Madler L, Keller AA, Cherr GN (2016) Developmental effects of two different copper oxide nanomaterials in sea urchin (*Lytechinus pictus*) embryos. Nanotoxicology 10:671–679

Towett EK, Shepherd KD, Drake BL (2016) Plant elemental composition and portable X-ray fluorescence (pXRF) spectroscopy: quantification under different analytical parameters X-ray. Spectrometry 45:117–124

Trujillo-Reyes J, Majumdar S, Botez CE, Peralta-Videa JR, Gardea-Torresdey JL (2014) Exposure studies of core-shell Fe/Fe(3)O(4) and Cu/CuO NPs to lettuce (*Lactuca sativa*) plants: are they a potential physiological and nutritional hazard? J Hazard Mater 267:255–263

Van NL, Ma C, Shang J, Rui Y, Liu S, Xing B (2016) Effects of CuO nanoparticles on insecticidal activity and phytotoxicity in conventional and transgenic cotton. Chemosphere 144:661–670

Vance ME, Kuiken T, Vejerano EP, McGinnis SP, Hochella MF Jr, Rejeski D, Hull MS (2015) Nanotechnology in the real world: redeveloping the nanomaterial consumer products inventory. Beilstein J Nanotechnol 6:1769–1780

VandeVoort A, Arai Y (2018) Macroscopic observation of soil nitrification kinetics impacted by copper nanoparticles: implications for micronutrient nanofertilizer. Nanomaterials (Basel) 8 (11):927

Vicario-Pares U, Lacave JM, Reip P, Cajaraville MP, Orbea A (2018) Cellular and molecular responses of adult zebrafish after exposure to CuO nanoparticles or ionic copper. Ecotoxicology 27:89–101

Vishnevetsky M, Ovadis M, Vainstein A (1999) Carotenoid sequestration in plants: the role of carotenoid-associated proteins. Trends Plant Sci 4:232–235

Wang Z, Li J, Zhao J, Xing B (2011) Toxicity and internalization of CuO nanoparticles to prokaryotic alga *Microcystis aeruginosa* as affected by dissolved organic matter. Environ Sci Technol 45:6032–6040

Wang Z, Xie X, Zhao J, Liu X, Feng W, White JC, Xing B (2012) Xylem- and phloem-based transport of CuO nanoparticles in maize (*Zea mays* L.). Environ Sci Technol 46:4434–4441

Wang Z, von dem Bussche A, Kabadi PK, Kane AB, Hurt RH (2013) Biological and environmental transformations of copper-based nanomaterials. ACS Nano 7:8715–8727

Wang T, Long X, Cheng Y, Liu Z, Yan S (2014) The potential toxicity of copper nanoparticles and copper sulphate on juvenile Epinephelus coioides. Aquat Toxicol 152:96–104

Wang Z, Xu L, Zhao J, Wang X, White JC, Xing B (2016) CuO nanoparticle interaction with *Arabidopsis thaliana*: toxicity, parent-progeny transfer, and gene expression. Environ Sci Technol 50:6008–6016

Wang S, Li Z, Gao M, She Z, Ma B, Guo L, Zheng D, Zhao Y, Jin C, Wang X, Gao F (2017) Long-term effects of cupric oxide nanoparticles (CuO NPs) on the performance, microbial community and enzymatic activity of activated sludge in a sequencing batch reactor. J Environ Manage 187:330–339

Woo-Mi L, Youn-Joo A, Hyeon Y, Hee-Seok K (2008) Toxicity and bioavailability of copper nanoparticles to the terrestrial plants mung bean (*Phaseolus radiatus*) and wheat (*Triticum aestivum*): plant agar test for water-insoluble nanoparticles. Environ Toxicol Chem 27:1915–1921

Xiong ZT, Wang H (2005) Copper toxicity and bioaccumulation in Chinese cabbage (*Brassica pekinensis* Rupr.). Environ Toxicol 20:188–194

Xiong T, Dumat C, Dappe V, Vezin H, Schreck E, Shahid M, Pierart A, Sobanska S (2017) Copper oxide nanoparticle foliar uptake, phytotoxicity, and consequences for sustainable urban agriculture. Environ Sci Technol 51:5242–5251

Xu C, Peng C, Sun L, Zhang S, Huang H, Chen Y, Shi J (2015) Distinctive effects of TiO_2 and CuO nanoparticles on soil microbes and their community structures in flooded paddy soil. Soil Biol Biochem 86:24–33

You T, Liu D, Chen J, Yang Z, Dou R, Gao X, Wang L (2017) Effects of metal oxide nanoparticles on soil enzyme activities and bacterial communities in two different soil types. J Soil Sediment 18:211–221

Yruela I (2009) Copper in plants: acquisition, transport and interactions. Funct Plant Biol 36:409–430

Yu S, Liu J, Yin Y, Shen M (2018) Interactions between engineered nanoparticles and dissolved organic matter: a review on mechanisms and environmental effects. J Environ Sci (China) 63:198–217

Yuan J, He A, Huang S, Hua J, Sheng GD (2016) Internalization and phytotoxic effects of CuO nanoparticles in *Arabidopsis thaliana* as revealed by fatty acid profiles. Environ Sci Technol 50:10437–10447

Zakharova OV et al (2018) Biological effects of freshly prepared and 24-h aqueous dispersions of copper and copper oxide nanoparticles on *E. coli* bacteria. Nanotechnol Russ 13(3–4):173–181

Zhang Z, Ke M, Qu Q, Peijnenburg W, Lu T, Zhang Q, Ye Y, Xu P, Du B, Sun L, Qian H (2018) Impact of copper nanoparticles and ionic copper exposure on wheat (*Triticum aestivum* L.) root morphology and antioxidant response. Environ Pollut 239:689–697

Zhao L, Huang Y, Zhou H, Adeleye AS, Wang H, Ortiz C, Mazer SJ, Keller AA (2016a) GC-TOF-MS based metabolomics and ICP-MS based metallomics of cucumber (*Cucumis sativus*) fruits reveal alteration of metabolites profile and biological pathway disruption induced by nano copper. Environ Sci-Nano 3:1114–1123

Zhao L, Ortiz C, Adeleye AS, Hu Q, Zhou H, Huang Y, Keller AA (2016b) Metabolomics to detect response of lettuce (*Lactuca sativa*) to Cu(OH)2 nanopesticides: oxidative stress response and detoxification mechanisms. Environ Sci Technol 50:9697–9707

Zhao J, Ren W, Dai Y, Liu L, Wang Z, Yu X, Zhang J, Wang X, Xing B (2017a) Uptake, distribution, and transformation of CuO NPs in a floating plant *Eichhornia crassipes* and related stomatal responses. Environ Sci Tech 51:7686–7695

Zhao L, Hu J, Huang Y, Wang H, Adeleye A, Ortiz C, Keller AA (2017b) NMR and GC-MS based metabolomics reveal nano-Cu altered cucumber (*Cucumis sativus*) fruit nutritional supply. Plant Physiol Biochem 110:138–146

Zhu Y, Xu J, Lu T, Zhang M, Ke M, Fu Z, Pan X, Qian H (2017) A comparison of the effects of copper nanoparticles and copper sulfate on *Phaeodactylum tricornutum* physiology and transcription. Environ Toxicol Pharmacol 56:43–49

Zuverza-Mena N, Medina-Velo IA, Barrios AC, Tan W, Peralta-Videa JR, Gardea-Torresdey JL (2015) Copper nanoparticles/compounds impact agronomic and physiological parameters in cilantro (*Coriandrum sativum*). Environ Sci Process Impacts 17:1783–1793

Systematic Review of the Literature of Factors Affecting the Exposure and the Levels of Lead in Human Breast Milk

Abha Cherkani-Hassani, Imane Ghanname, Aida Maria Benitez-Rexach, and Nezha Mouane

Contents

A. Cherkani-Hassani (✉)
Unit of Training and Research in Nutrition and Food Sciences, Faculty of Medicine and Pharmacy, University Mohammed V, Rabat, Morocco

Center of Doctoral Studies in Life Sciences and Health CEDoc-SVS, Faculty of Medicine and Pharmacy, University Mohammed V, Rabat, Morocco
e-mail: abha.hassani@um5s.net.ma; abha.cher@gmail.com

I. Ghanname
Team Research of Pharmacoeconomics and Pharmacoepidemiology, Laboratory of Pharmacology and Toxicology, Faculty of Medicine and Pharmacy, University Mohammed V, Rabat, Morocco

Faculty of Health Sciences, International University of Casablanca, Casablanca, Morocco
e-mail: imane.ghanname@gmail.com

A. M. Benitez-Rexach
Walden University, Minneapolis, MN, USA
e-mail: a.benitez-rexach@outlook.com

N. Mouane
Unit of Training and Research in Nutrition and Food Sciences, Faculty of Medicine and Pharmacy, University Mohammed V, Rabat, Morocco

Department of Pediatric Hepatology Gastroenterology and Nutrition-P III, Rabat Children's Hospital, Rabat, Morocco
e-mail: nezhamouane@hotmail.com

© Springer Nature Switzerland AG 2019

P. de Voogt (ed.), *Reviews of Environmental Contamination and Toxicology, Volume 252*, Reviews of Environmental Contamination and Toxicology Volume 252,
https://doi.org/10.1007/398_2019_32

1 Introduction

Human milk is, without question, the best source of nutrition for infants. However, maternal milk may sometimes contain chemical contaminants, which could have adverse effects on neonates or nursing infants. Lead (Pb) is of considerable interest due to its toxicity and widespread use (Järup 2003). The nervous system is the principal target organ for Pb, although it may adversely affect most organs in the body. Lead accumulates in the human body through three compartments of kinetic importance: blood, soft tissue, and bone (Marcus 1985).

Lead is ubiquitous in the environment, but human activities can substantially raise its dissemination and background levels. Therefore, Pb exposure to human populations is through the food chain, drinkable water, and air, as it was estimated that food and water consumption accounted for approximately 40% of Pb acquisition for non-occupationally exposed groups (WHO 1996).

Breast milk has been suggested as a significant potential source of lead exposure to nursing infants (Silbergeld 1991). The half-life of lead in human milk is approximately 3 months (Anderson and Wolff 2000), and the range of the standard values reported by the World Health Organization is 2–5 μg/L (WHO 1989).

Much of the lead in breast milk does not come from the mothers' exposures during lactation. Instead, it comes from Pb stored in the bones (Gulson et al. 1998b). The process that mobilizes bone calcium during pregnancy and lactation can also assemble Pb storage, posing an exposure risk to the newborn, and so, sufficient calcium (Ca) intake during pregnancy and lactation reduces the extraction of lead from the bone (Byczkowski et al. 1994; Gulson et al. 1998b; Silbergeld et al. 1988).

Inorganic lead is classified as a possible human carcinogen (group 2A) by IARC, and organic lead compounds are "not classifiable to its carcinogenicity to humans" (Group 3) (IARC 2006).

A large number of published studies have investigated the levels of lead in human breast milk. The first study dates from 1974 (Dillon et al. 1974), but only a few of them searched and explored the factors of maternal exposure that affected the level of Pb in human milk among nursing mothers, and these were selected in this review.

The aim of this review was to explore the published literature, monitoring studies that identified common factors affecting the level of lead in human breast milk.

2 Review Method

2.1 Search Strategy

The search strategy and review processes followed the general guidelines of the Preferred Reporting Items for Systematic Reviews and Meta-Analyses (PRISMA) (Moher et al. 2009).

Online searched databases included PubMed, Science Direct, and Google Scholar for studies investigating the related factors affecting the levels of lead in human breast milk among lactating mothers around the world. The combination keywords were the following: lead, heavy metals, trace elements, toxic metals, human milk, and breast milk, as well as affecting factors, to select relevant studies. Reference lists of identified papers were also checked for suitable publications. There were no restrictions on the beginning date of this search, but the end was 2017, and the studies must be in the English language. Also, the major component of the review was that the included papers must correlate at least one or many factors with the level of lead in breast milk.

2.2 Studies Selection

The study reached over 740 articles in the initial search. After excluding duplicates found within the research, approximately 260 papers were retained from the literature. Furthermore, after screening exclusion criteria, 120 articles were retrieved for eligibility. Studies not done in English language or reported data from books, reports, theses, conferences, symposia, reviews, and poster abstracts from congresses were excluded, as well as articles conducted to develop analytical methods or validate them. Lastly, full-text articles were then reviewed to inquire if at least one or more factors were evaluated to affect the level of Pb in HBM. But about 40 reported data only on Pb concentrations in HBM without assessing any factors of exposure. Accordingly, 49 papers were identified and included conclusions relating to the assessed factors that may or not may affect the level of Pb in HBM.

Figure 1 summarizes strategies for searching, screening, eligibility, and including studies (Flow diagram of literature search process).

2.3 Data Extraction

From the 49 papers that met the selection criteria, a data retrieval form was used to record results. The criteria used were the following: general information (authors, date, country), specific information (aims and design of the study, sample size, lactation stage, study population or area of residence of the measured concentration,

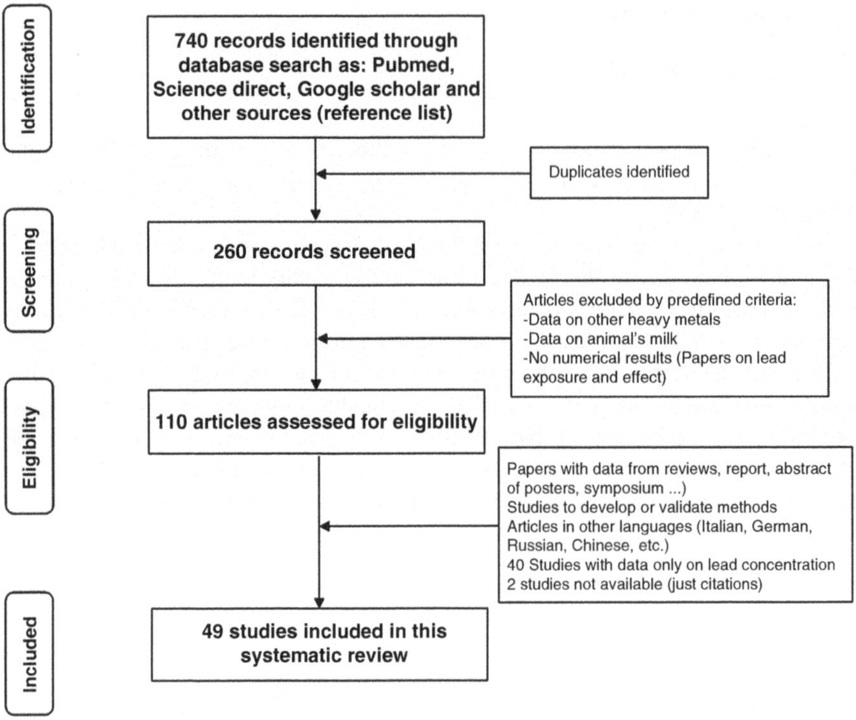

Fig. 1 Flow diagram of the literature search process

maternal age, lead concentration (mean ± SD, median, geometric mean, and range)), relevant findings (the associated factors), the statistical tests used, and author's conclusions.

3 Results and Discussion

3.1 Search Results

From the reviewed literature, 49 papers rated and explored the effect of one or several factors on the level of Pb in HBM from 28 countries. The studies were carried out over an extended time frame from 1983 to 2017 and through a large geographical area: Europe, Asia, America, and Africa, regarding more than 5,000 subjects. In fact, 21 studies were conducted in Europe and from the following countries: Turkey (four studies), Portugal (three studies), Germany, Sweden, Croatia, Greece, Austria, Russia, Italy, Slovakia, Spain, and Poland (one study from each country).

Furthermore, more studies were explored from Asian and Middle East countries. Two studies were reviewed from India and Taiwan; one study from Korea and Malaysia, respectively; three studies from Saudi Arabia; as well as one study from the United Arab Emirates, Egypt, Iran, Iraq, and Palestine each, for a total of 14.

Moreover, a final group of studies was explored from the Americas and Africa: three studies each from the United States, Brazil, and Mexico, two studies from Canada, and one study each from Nigeria and Ghana, for a total of 13.

3.2 Design and Characteristics of Studies

The literary review displayed a majority of cross-sectional studies ($n = 38$) with few longitudinal surveys ($n = 10$) and only one prospective cohort study (Park et al. 2017). Precisely, the factors chosen for the reviewed studies were (a) the potential changes in Pb levels during the lactation stage (colostrum to transition milk to mature milk) and (b) place of residence (urban/rural/industrial areas). Other factors chosen were individual maternal choices: age, parity, dietary habits, smoking habits, and the level of Pb in HBM. The main features of the included studies are in Table 1.

The results of the 49 studies that explored the factors affecting the level of lead in human breast milk were categorized into four sections:

1. Physiological and lactation factors: age, weight, height, BMI, bone, skin color, lactation stage, parity, gestational age, parturition, duration of the previous breastfeeding, number of children, milk composition of the essential minerals, nutritional status, history of anemia
2. External factors: dietary intakes, smoking habits and exposure, drinking alcohol, kind of utensils used, place of residence (area, duration, house characteristics), mineral and vitamin supplementation, amalgam fillings, use of kohl, exposition to heavy traffic and insecticide
3. Socioeconomic factors: maternal level of education and work status, family incomes, husband's occupation
4. Newborn factors: gender, birth and current weight and length, and head circumference

The summary of the studies with associated and not associated factors as well as other factors showing differences of Pb level in HBM is shown in Tables 1 and 2.

3.3 Statistical Methodology

The studies were further examined for statistical correlation of Pb levels measured in HBM with several factors, including maternal, physiological, external, socioeconomic, and newborn factors. Many tests were used, such as Student's *t*-test, ANOVA, Mann-Whitney test, Wilcoxon test, Kruskal-Wallis test, Pearson test, Spearman's test, the Nagelkerke R^2, and univariate and multivariate regression. The observed differences were statistically significant at $p < 0.05$.

Table 1 Summary of literature studies on lead levels and factors in human breast milk

Authors, country	Sample size	Lactation stage	Age of mothers mean ± SD [range]	Note (population/ area of residence/ type of milk (N))	Lead concentration µg/L Mean ± SD/ geometric mean[GM]	Lead concentration µg/L Median (range)	Factors tested	Affecting factors found
Huat et al. (1983), Malaysia	180	Various		Urban (89) Rural (91)	25.3±11.3 21.1±9.8	(2.1–58.9) (4.4–53.8)	Place of residence	*Place of residence:* Urban samples were significantly higher than rural ($p > 0.001$)
Rockway et al. (1984), USA (LONGI)	39 W 62	1–22 Mpp	[22–47]	Arizona Urban and rural women	2.8±1.6	(0.9–10)	Lactation stage Place of residence	No differences were found
Rabinowitz et al. (1985), USA	100	1 & 6 Mpp	30	Boston	1.7±0.2	(0–7.2)	Lactation stage	*Lactation stage*: Pb in samples collected at 1 month differs minimally from those collected at 6 months
Sternowsky and Wessolowski (1985); Germany (LONGI)	20 W	2–90 Dpp	>20	Colostrum Mature milk 10 city 10 rural	15.5 ±6.1 9.1 ±2.5 13.2±5.5 9.1± 2.5		Place of residence Lactation stage Maternal age	*Place of residence:* Higher values in urban population compared to rural population. Urban women have mean levels seemed to increase with duration of lactation, while in the rural area, they decreased *Lactation stage:* Higher values in colostrum compared to mature milk *Maternal age:* Women over 30 years of age excreted a significantly larger amount of lead than those under 30
ONG et al. (1985), Malaysia	114 W 9 W (LONGI)	3–5 Dpp 3 Dpp to 1 Mpp		Colostrum (114) Transition and mature milk (53)	47.84±12.48 43.68	24.96–106.08	Lactation stage	*Lactation stage*: Pb did not vary significantly from 3 days to over 1 month postpartum
Dabeka et al. (1986); Canada	210				1.04 0.566^{GM}	0.55 (<0.05–15.8)	House age, exposition to heavy traffic, dietary habits (coffee consumption)	*House age:* Variable most strongly associated with lead levels in milk ($p < 0.001$). Occupants of older houses had, on average, increased lead levels, particularly those living in houses

								that were more than 30 years old *Exposition to heavy traffic:* Exposure of the mother to heavy traffic for more than 5 years was the variable with the next highest degree of association with milk lead levels ($p = 0.011$) *Dietary habits (coffee consumption):* Increased lead levels were noted among individuals drinking 3 cups and 4 or more cups ($p = 0.034$)
Namihira et al. (1993), Mexico	35 W	1–27 Mpp	24.7 [15–39]	3 smelter areas	61.8 24.7^{GM}	(9.2–351)	Parity, stage of lactation, number of children, years of residence in the areas	*Number of children*: Number of children stratified by 1 or 2 and 3 or 4 children with $p < 0.001$
Younes et al. (1995), Saudi Arabia	47	1–60 Dpp		Overall According to: Maternal age Lactation stage	7.68±4.2 5.15±1.4 to 13.15± 6.5 6..6 ± 3.8 to 8.86 ± 5.3	3.18–25	Maternal age Lactation stage Place of residence Dietary and culinary habits	*Maternal age:* The lowest lead concentration was found in milk samples obtained from mothers age 20 years or less and increased progressively with age and found in significantly high levels in mothers age 36 years or more ($p \leq 0.05$) *Lactation stage:* No significant differences between lead concentrations in samples obtained from right or left breasts, and similarly, there were no significant differences in lead concentrations in milk samples in relation to the length of period of lactation *Place of residence:* Samples obtained from mothers residing near industrial areas or highways showed higher lead concentration than from those living in remote areas with reduced exposure

(continued)

Table 1 (continued)

Authors, country	Sample size	Lactation stage	Age of mothers mean ± SD [range]	Note (population/ area of residence/ type of milk (N))	Lead concentration μg/L		Factors tested	Affecting factors found
					Mean ± SD/ geometric mean[GM]	Median (range)		
								Kind of utensils used: Samples obtained from mothers using copper casserole coated with white (rich in lead) inner coat and eating food material preserved for long periods in metal containers showed higher lead concentration
Hallén et al. (1995); Sweden	74	6 Wpp	29 [17–45]	Smelter area (39) Control area (35)	0.9 ± 0.4 0.5 ± 0.3		Place of residence Dietary habits (wine consumption) Smoking habits	*Place of residence* had a strong impact on the milk lead concentration ($p < 0.001$)
Saleh et al. (1996), Egypt	120		25 [20–40]	20 different cities	30.6	(0–158)	Place of residence	*Place of residence*: Lead levels in mother's milk from Alexandria, Assiut, and Cairo were significantly higher than in the other cities
Frković et al. (1997); Croatia	29	2–12 Dpp	28.9 [17–45]	Overall According to: *Maternal age*: Mothers ≤25 years (10) Mothers >25 years (19) *Parity* Primiparous (14) Multiparous (15) *Place of residence* Rijeka (13) Region (16) *Smoking habits* Smokers (7) Non-smokers (22)	7.3±8.3 10.4±13.1 5.7±3.5 5.8±4.2 8.7±10.8 10.6±10.5 4.7±4.8 5.7±6.7 7.9±8.8	5 (0.3–44.0) (0.5–44) (0.3–12) (0.5–12.4) (0.3–44) (2.9–44) (0.3–19) (0.5–19) (0.3–44)	Maternal age Parity, place of residence, smoking habits	*Maternal age:* Lead in women aged ≤25 years was higher compared with group of >25 years, but the difference was not statistically significant ($p > 0.05$) *Parity:* Of 29 women studied, 14 were primiparous and had somewhat lower concentrations of lead in breast milk compared to women with two or more deliveries, but the difference is not significant *Place of residence:* Mothers living within city area had higher concentration of lead in breast milk ($p < 0.01$), compared to those living in smaller towns and settlements within the region

								Smoking habit: It does not have influence on the lead content of breast milk ($p > 0.05$)
Vavilis et al. (1997), Greece	91	Colostrum		Urban mothers (51) Rural mothers (40)	90±29 84±24		Place of residence	*Place of residence*: Difference was not statistically significant
Klopov (1997), Russian Arctic	42	1–3 Dpp		Norilsk Salekhard	26.67±2.64 22.5±3.56		Place of residence	*Place of residence*
Gulson et al. (1998a), Austria, LONGI	9	10–180 Dpp			0.73±0.70	(0.09–3.1)	Dietary habits Bone	The major sources of lead in breast milk are from the maternal *bone and diet*
Krachler et al. (1998); Austria (LONGI)	55 46 W	1–293 Dpp		Overall 1–3 Dpp (13) 4–17 Dpp (18) 42–60 Dpp (8) 66–90 Dpp (8) 97–293 Dpp (8)	1.5 2.3 2.7 2.4 1.5 0.9	2.1 (<0.12–9.9) 1 (0.2–5.6) 2 (<0.12–8.7) 1.5 (<0.12–9.9) 1.1 (<0.12–4.6) <0.12 (<0.12–4.7)	Lactation stage	*Lactation stage:* Concentrations for Pb in the milk samples had a statistically significant trend ($p < 0.001$) toward lower concentrations during lactation
Friel et al. (1999); Canada (LONGI)	43	1–12 Wpp	20–35	*PRT:* *<37 Week (24)* ND for all weeks except 7th Wpp *FT: 39–41 Week (19)* 2 Wpp 3 Wpp 4 Wpp 5 Wpp 6 Wpp 7 Wpp 8 Wpp 12 Wpp		1 0.5 2 1 4 3 4 3 2	Lactation stage Gestational age	*Lactation stage:* For the FT group, Pb showed an overall, although fluctuating, increase in median concentration until week 7, after which time the concentration decreased. For the PRT group, there was no increase or decrease *Gestational age:* Showed no significant differences but median difference between preterm and full-term milk: lower Pb in PRT milk than FT
Vanderjagt et al. (2001), Nigeria	34	2–24 Wpp	26.0 ± 7.5 [15–45]	Nomadic women	102	67 (<46–1,300)	Maternal age BMI Parity	No statistical difference
Gundacker et al. (2002), Austria	138		29±5	Overall According to: *Residence:* Countryside: Tullin (48)	1.63±1.66 1.22±0.92 2.48±2.39 1.29±1.12 0.80±0.80		Place of residence Dietary habits (cereals and fish consumption) Maternal weight	*Place of residence ($p < 0.05$)* *Fish consumption ($p < 0.05$)* *Smoking habits ($p < 0.05$)*

(continued)

Table 1 (continued)

Authors, country	Sample size	Lactation stage	Age of mothers mean ± SD [range]	Note (population/ area of residence/ type of milk (N))	Lead concentration μg/L: Mean ± SD/ geometric mean[GM]	Lead concentration μg/L: Median (range)	Factors tested	Affecting factors found
				Industrial: Linz (37) Urban :Vienna (27) *Fish consumption* Never (8) Occasionally (87) Frequently (17) *Smoking habits* Never (86) Past smokers (12) Current smoker (14)	1.69±1.75 1.82±1.31 1.57±1.67 1.15±0.94 2.40±2.0		Smoking habits Prematurity Vitamin supplementation	
Sowers et al. (2002), USA, New Jersey (LONGI)	15 W	1.5–12 Mpp	23.7±1.4	1.5 Mpp 3 Mpp 6 Mpp 12 Mpp	6.1±1 5.6±1.1 5.9±1 4.3±1.6		Lactation stage	*Lactation stage:* At each time point, concentrations were significantly different, but there was no significant linear trend across time ($p = 0.45$)
Al-Saleh et al. (2003), Saudi Arabia	362		29.81 [15–51]	*Overall* Riyadh (168) ALEhssa (194)	31.67±45.66 25.164±38.771 37.306±50.305	20.998 (<1.325–489.83) 16.397 (<1.325–352.9) 26.889 (<1.325–489.83)	Place of residence Lactation stage, current and birth weight, mother's age, mother's weight, parity, average duration of lactation for all births, education level, job, the family's income, and dietary habits (consumption of fish)	*Place of residence:* Higher significantly in ALEhssa than Riyadh *Average duration of lactation:* Significant *Dietary habits (fish consumption)*: Mothers who did not eat fish had higher Pb in breast milk with statistical significance
da Silva Anastácio et al. (2004), Brazil	38	20–102 Dpp	27.7±6.5 [19–43]	Low environmental lead exposure from Rio de Janeiro	2.8±2.5 1.2 [GM]	(<0.1–11.9)	Dietary intake, milk composition of the essential minerals	*Calcium intakes*: Concentrations of lead and calcium in whole milk were positively associated ($r = 0.56$; $p = 0.005$)

Turconi et al. (2004); Italy	143	3–4 Dpp	[18–42]	Northern Italy		7.75 (2–216)	Place of residence, kind of job, dietary and smoking habits	No statistical significant differences
Ettinger et al. (2004), Mexico	310	1 Mpp	24.4±5 [14–43]	Overall *Mother's age* <20 (49) 20–30 (220) >30 (39) *Current use of lead-glazed ceramics* Yes (126) No (184)	1.20±0.83 1.44±1.19 1.11±0.65 1.53±1.17 1.26±1.03	1.1 (0.21–8.02)	Maternal age Parity Previous lactation Years living in Mexico City Daily intake of calcium *Kind of utensils used* (use of lead-glazed ceramics)	*Maternal stage:* Older women had breast milk lead levels lower than those in the younger age categories *Parity:* Mean breast milk lead levels were lower for primiparous women than for women with one previous pregnancy but lower still for women with two or more previous pregnancies, but differences were not statistically significant *Previous lactation*: For women who had breastfed previously for 12 months or more (1.31 ±1.27), breast milk lead levels were slightly lower than for those women who had breastfed previously for less than 12 months (1.38 ± 0.90) or not at all (1.38 ± 1.22), but differences were also not statistically significant *Calcium intakes*: Women with daily intake of calcium <1,000 mg had higher Pb (1.45 ± 1.19) than those who received more adequate amounts of calcium (≥1,000 mg) in their diets (1.30 ± 1.02), but no significant difference *Kind of utensils used (use of lead-glazed ceramics)*: ($p = 0.03$)
Ettinger et al. (2005), Mexico (LONGI)	310	1,4, and 7 Mpp		1 Mpp (310) 4 Mpp (224) 7 Mpp (195)	1.4±1.1 1.2±1.0 0.9±0.8	1 (0.2–8.0) 0.9 (0.2–6.8) 0.7 (0.2–4.8)	Maternal age Lactation stage, breastfeeding practice Parity	*Lactation stage:* Lead levels showed a significant decreasing trend over the course of lactation ($p < 0.00001$) by 4 and 7 months postpartum

(continued)

Table 1 (continued)

Authors, country	Sample size	Lactation stage	Age of mothers mean ± SD [range]	Note (population/ area of residence/ type of milk (N))	Lead concentration μg/L		Factors tested	Affecting factors found
					Mean ± SD/ geometric mean[GM]	Median (range)		
							History of previous lactation Past use of lead-glazed ceramics Dietary calcium intake	The levels of lead in breast milk at 1 month postpartum were significantly correlated with breast milk lead at 4 months ($r = 0.21$, $p = 0.006$) and 7 months ($r = 0.29$, $p = 0.0003$) postpartum *Kind of utensils used (use of lead-glazed ceramics)*: Current use of lead-glazed ceramics was associated with higher breast milk lead levels at 1 month postpartum ($p = 0.04$). However, by 4 and 7 months postpartum, no significant differences remained were also significantly correlated with each other ($r = 0.30$, $p < 0.0001$)
Ursinyova and Masanova (2005), Slovakia	158	4 Dpp	25.6 [18–39]	8 different locations of the Slovak Republic *Overall* *Maternal age* Mothers ≤26 years (102) Mothers >26 years (55) *Parity* Primiparous (72) Multiparous (83) *Mother's smoking BP* Yes (36) No (123)	4.7 4.29±3.49 5.38±5.04 5.09±5.13 4.33±3.13 5.13±4.17 4.51±4.13	3.4 (ND–24.4)	Place of residence, maternal age, parity Number of mother's tooth fillings, newborn's gender, and birth weight Smoking habits	*Maternal age*: Mean milk Pb in women aged ≤26 years were lower compared to the group with >26 years of age, but differences were not statistically significant ($p = 0.11$) *Parity:* Primiparous women had somewhat higher levels of Pb in milk, but differences were not statistically significant *Smoking habit:* Lead level in milk was somewhat higher in mothers smoking before pregnancy than in non-smokers, but a difference was not significant

Leotsinidis, et al. (2005), Greece	180	3 Dpp 14 Dpp	25 (16–39)	Colostrum (180) Transitory milk (95)	0.48±0.6 0.15±0.25	0.44 (ND-2.3) ND (ND-0.94)	Place of residence Education level Dietary and smoking habits Lactation stage Maternal age and weight, neonate weight, length and head circumference, occupation, parturition, parity and vitamin supplementation	*Lactation stage:* A decline of Pb concentrations was observed in transitory milk in comparison to the colostrums at a statistically significant level *Place of residence:* Higher values in urban area $p = 0.035$ *Education level:* Elementary and high school versus higher education ($p = 0.015$) *Dietary habits:* Consumption of red meat seems to be a protective factor for lead (OR 0.168, 95% CI 0.053–0.534). The lead is influenced positively by cheese (OR 3.27, 95% CI 1.14–9.37) and rice (OR 4.38, 95% CI 1.33–14.37)
Sharma and Pervez (2005), India	110	<1 Wpp	[20–45]	Workers (40): 20–25 years 40–45 years Nonworkers (45) (Township residents) Uncontaminated (35)	3.6± 2.0 22.3± 18.5 0.6± 0.2 to 4.2±3.6 <LOD to 0.7±0.3		Maternal age Place of residence	*Maternal age:* The accumulation is higher in older age group. Good positive correlation between the age group of the subjects and accumulation of toxic metals are found *Place of residence*
Chien et al. (2006), Taiwan	72	Colostrum	30.8±4.2 31.5±4.0	Overall Mothers who consumed traditional Chinese herbs (35) Control group (37)	$7.86^{GM}±8.24$ $8.59^{GM}±10.95$ $6.84^{GM}±2.68$		Dietary habits Lactation stage Age, height, weight, and body mass index (before and during pregnancy), neonatal body weight, parity, drinking and smoking habits	*Dietary habit (traditional Chinese herbs consumption)*: In the consumption group, lead is statistically significantly higher than for the control group ($p < 0.05$) *Lactation stage:* Lead concentration in breast milk decreases as a function of lactation time

(continued)

Table 1 (continued)

Authors, country	Sample size	Lactation stage	Age of mothers mean ± SD [range]	Note (population/ area of residence/ type of milk (N))	Lead concentration μg/L Mean ± SD/ geometric mean^{GM}	Median (range)	Factors tested	Affecting factors found
Kinsara et al. (2006); Saudi Arabia (LONGI)	156 11 W	1 to 15 Dpp	25	Jeddah city Overall (1–15 Dpp) Morning (1–7 Dpp) Afternoon (1–7 Dpp)	2.7 1.47±0.01 1.76±0.1		Lactation stage (morning and afternoon)	*Lactation stage (morning and afternoon):* The concentrations of the elements in milk samples collected in the afternoon are higher than those in the morning *Stage of lactation*: There is an apparent decline in the mean concentration levels as the stage of lactation progressed
Abballe et al. (2008), Italy	39	4–8 Wpp	[21–40]	Rome (10) Venice LC (10) MC (13) HC (6)	0.849 0.971 1.07 0.979		Place of residence Dietary habits (fish consumption)	Not affect
Abdulrazzaq et al. (2008), UAE	205	4–80 Wpp	[18–50]	Al Ain city 68 mothers were from the UAE and 124 were from other nationalities	0.019±0.055	0.4 (0.00–0.55)	Place of residence Maternal age weight and height, age and sex of the child, duration of residence in UAE (5 years), mother's education, job, smoking habits	No association
Almeida et al. (2008), Portugal	34	2 Dpp and 1 Mpp		Colostrum (34) Mature milk (19)	1.55±1.38 0.94±1.05	1.17 (0.06–5.43) 0.53 (0.07–4.03)	Lactation stage	*Lactation stage:* Pb was lower in mature milk, but the decrease did not reach statistical significance
Rahimi et al. (2009); Iran	44 W	1–6 Wpp		*Industrial city (overall)* *Mother's age* <30 (22) ≥30 (22) *Parity* Primiparous (22) Multiparous (22)	10.39 ± 4.72 8.92 ± 3.28 11.86 ± 5.51 8.71 ± 4.11 12.07 ± 4.78	(3.18–24.67) (3.18–14.27) (4.70–24.67) (3.18–19.88) (4.67–19.88)	Maternal age Parity Smoking habits	*Maternal age:* Pb higher in old mothers ($p < 0.05$) *Parity:* Pb higher in multiparous mothers ($p < 0.05$)

Koka et al. (2011), Ghana	54		[18–39]	*Kind of water* Tap water (34) Borehole water (20) *Smoking habits* Passively exposed (16) None exposed (38)	*24±5 *20±6 *26±3 *20±2	(22–32) (16–22) (24–32) (16–24)	*Dietary habits* (type of drinking water) Smoking habits	*Dietary habits* (type of drinking water): There was statistically a significant increase in lead levels in breast milk of mothers drinking tap water compared to mothers drinking borehole water ($p < 0.05$) *Smoking habits:* A significant increase of Pb levels in breast milk of women passively exposed to smoking compared to none exposed ($p < 0.05$)
Koyashiki et al. (2010), Brazil	92	45–210 Dpp	28.1 [16–42]	From breast milk bank	2.9±1.1	3 (1–8)	Maternal age, smoking habits Dietary habits (milk consumption) Lactation stage Number of children, economic class, living near contamination source, living with people exposed to lead, skin color	*Skin color:* Lower in white skin than in not white ($p = 0.023$)
Cinar et al. (2011), Turkey	75	1 Mpp		Industrial area (20) City center (15) Rural area (21)	282±155 176±151 303±199	(0.00–546) (0.00–450) (0.00–628)	Place of residence	*Place of residence*: Pb in rural areas was highest Statistically significant difference exists ($p < 0.05$)
Isaac et al. (2012), India	50	0.5–16 Wpp	[25–44]	Non industrial area (25) Industrial area (25)	13.21±5.2 21.5±4.5	13 (9–21) 18 (15–25.5)	Place of residence	*Place of residence*: Higher in industrial area. *Lactation stage*: A significant decline in the Pb levels was noticed in the breast milk of mothers of the nonindustrial area in comparison with samples from mothers of industrial area The lactating mothers from industrial area have lower IQ

(continued)

Table 1 (continued)

Authors, country	Sample size	Lactation stage	Age of mothers mean ± SD [range]	Note (population/ area of residence/ type of milk (N))	Lead concentration µg/L Mean ± SD/ geometric mean[GM]	Median (range)	Factors tested	Affecting factors found
								(intelligence quotient) levels compared to mothers from nonindustrial area
Orun et al. (2011); Turkey	144	2 Mpp	25 [17–41]	Suburban areas of Ankara (overall) History of anemia Yes (45) No (32)		20.59 (<0.2–1,515) 21.1 17.9	The maternal age, weight, gestational age, education level, family monthly income, place of residence, parity, gender of the newborn, the employment status, smoking habits and birth interval, vitamin and iron intake	*History of anemia* :The mothers with a history of anemia at any time had higher breast milk Pb levels than those without anemia ($p = 0.0052$)
Garcia-Esquinas et al. (2011), Spain	100	3 Wpp	30.6	Madrid city	15.56[GM]	16.64	Sociodemographic variables (age, birth place, education level, geographic area of residence) or variables related with previous pregnancies and lactations or environmental factors (tobacco exposure and amalgam fillings)	*Dietary habits (potato consumption, caffeinated coffee beverages, and dairy products)*: Mean Pb levels increased with consumption of potatoes (p trend = 0.02) but showed an inverse association with ingestion of caffeinated coffee beverages (p trend = 0.03) and dairy products (p trend = 0.04) *Exposure to motor vehicle traffic density* :Displayed higher lead levels in breast milk
Nassir et al. (2013); Iraq	68	1–6 Wpp	26.1 [16–40]	Overall *Age of mothers:*	25.9±18.40 25.0±3.48 to		Maternal age BMI	*Maternal age* *BMI*

				16–30 31–40 *BMI* Normal weight Over weight Obesity *Palace of resi-dence:* Urban area Rural area Industrial area Norman area *Drinking water* Tap water River water Bottled water *Smoking:* Smokers Non-smokers	32.99±5.20 14.05±0.99 to 16.26±5.44 11.13±3.25 29.09±3.81 20.96±2.81 31.65±22.19 19.59±13.66 38.64±35.52 16.19±9.14 30.54±3.87 30.88±4.33 15.90±2.57 54.79±27.97 22.08±12.80		Place of residence Dietary habits (type of drinking water) Smoking habit	*Place of residence* *Dietary habits (type of drinking water)* *Smoking habit* All differences were significant ($p < 0.05$)
Marques et al. (2013), Brazil	82	0.5–12 Mpp 1–24 Mpp	[14–40] [16–40]	Smelter area (37) Rural fishing village (45)	12.64 ±8.16 4.3± 4.01	11.3 (0.9–29.4) 3.5 (0–16.2)	Place of residence, years of residence, maternal age Lactation stage, number of children	*Place of residence:* Level of BM-Pb concentration of urban exposed mothers was significantly higher ($p = 0.0000001$) than that of rural mothers *Years of residence:* Mothers also showed a statistically significant correlation between length *of residence and BM*-Pb concentration ($r = 0.6864$; $p = 0.0001$)
Matos et al. (2014); Portugal (LONGI)	155 31 W	1 to 16 Wpp	30.9 [21–39]	1 Wpp 4 Wpp 8 Wpp 12 Wpp 16 Wpp	0.156±0.62 0.146±0.67 0.066±0.67 0.18±0.71 0.26±0.68	0.11(−0.93–1.14) 0.17 (−0.91–1.59) 0.10 (−0.87–1.38) 0.11 (−0.85–1.76) 0.4 (−0.92–1.32)	Maternal age Lactation stage Parity Intake of mineral supplements	*Lactation stage:* The increase of Pb in mature milk did not reach statistical significance
Hamzaoglu et al. (2014), Turkey	53	1–4 Dpp	27 [19–36]	Industrial (36) Nonindustrial (16)		8.7 (<LOQ-96.0) 10.1 (<LOQ-26.7)	Place of residence	*Place of residence*
Winiarska-Mieczan (2014); Poland	320 W	1–12 Mpp	[20–40]	Lublin region (overall) *Mothers age*	6.331± 4.614 4.762 6.932	(0.486–12.01)	Maternal age Lactation stage Smoking stage	*Maternal age:* Most Pb was discovered in the breast milk of women aged 36 to 40 and 26 to

(continued)

Table 1 (continued)

Authors, country	Sample size	Lactation stage	Age of mothers mean ± SD [range]	Note (population/ area of residence/ type of milk (N))	Lead concentration μg/L Mean ± SD/ geometric mean[GM]	Median (range)	Factors tested	Affecting factors found
				20–25 26–30 31–35 36–40 *Lactation stage* 0–1 Mpp 2–3 Mpp 4–6 Mpp 7–12 Mpp *Smoking habits:* Smokers Non-smokers	6.221 7.411 2.7 3.3 3.0 2.0 9.5 6.5			30 (approx. 7–7.4 μg/L). The lowest content of Pb was characteristic of the milk of women aged 20 to 25 *Lactation stage:* The highest content of Pb was found in milk between 2 and 3 months of lactation, while it was definitely lower in milk between 7 and 12 months of lactation *Smoking habits:* Pb in the milk of women who smoked was significantly higher than in non-smokers
Chao et al. (2014); Taiwan (LONGI)	45 W	1–4 Dpp (C) 5–10 Dpp (T) 30–35 Dpp (M1) 60–65 Dpp (M2)	29.84 ±4.58 [22–39]	C T M1 M2	13.22 ±3.58 8.92± 2.60 11.72±2.58 2.93±1.70	(6.70–22.36) (3.52–14.71) (0.76–11.72) (0.45–7.80)	Maternal age Lactation stage Parity Smoking habits Mothers' level of education Exposition to insecticide	*Lactation stage:* Lead concentration was significantly higher ($p < 0.01$) in colostrums compared to transitional milk and mature milk *Maternal age:* The lead levels in breast milk were also higher in older mothers, and these differences were statistically significant *Parity*: Multiparous women had somewhat higher levels of lead in their milk *Mother's level of education*: The lead levels in breast milk were also higher in mothers with higher educational level, and these differences were statistically significant

Dursun et al. (2016), Turkey	107	3–10 Dpp	30.6±4.9 [21–46]		14.56±12.13	(<0.03–46.14)	Maternal age, working status, parity, gestational age, neonatal anthropometry and gender, maternal tobacco exposure during childhood and pregnancy, fish consumption and amalgam felling, exposition to insecticide, kind of utensils used (use of lead-glazed ceramics), lead in home plumbing and type of home heating system	*Exposition to insecticide*: Higher level of Pb in breast milk of exposed mothers than those not exposed to insecticide (17.03 ± 11.8 versus 12.31±12.05 μg/L, $p < 0.05$)
Letinić et al. (2016), Croatia	104 W	C, T, M	30±5	*Non-smokers* Colostrum (20) Transition milk (33) Mature milk (51) *Stopped smoking during pregnancy* Colostrum (5) Transition milk (11) Mature milk (15) *Continued smoking during pregnancy* Colostrum (9) Transition milk (14) Mature milk (20)		5(2.6–10) 3.4(1.9–5.6) 2.6(1.7–4.7) 5(4.5–6.6) 3.8(3.1–8.2) 4.0(2.9–7.2) 2.7(2.4–5.7) 8.5(3.9–12) 3.3(2.0–5.5)	Lactation stage and smoking habits	*Lactation stage* correlated to *smoking habits*: The concentrations of Pb were almost twofold higher in colostrum of non-smokers and about 2.5 times higher in transitional milk of smokers who continued smoking during pregnancy
Shawahna, et al. (2016), Palestine	89	15–210 Dpp	26.8±5.6	Nablus, Ramallah, and Jerusalem regions		4 (2–12)	Maternal age, number of children, lactation stage, place of	*Place of residence* ($p < 0.05$) *Residence in house with peeling or chipping paint* ($p < 0.05$)*or*

(continued)

Table 1 (continued)

Authors, country	Sample size	Lactation stage	Age of mothers mean ± SD [range]	Note (population/ area of residence/ type of milk (N))	Lead concentration µg/L Mean ± SD/ geometric mean^{GM}	Lead concentration µg/L Median (range)	Factors tested	Affecting factors found
							residence (city, village, or refugee camp), if the house paint was peeling or chipping, distance from a contamination source (paint shop, industrial area, and/or gas station), household income, educational level, employment status, husband's occupation, working in agriculture, use of cosmetics, eye kohl, hair dyes, and clay utensils. Smoking habits and alcohol consumption	*near paint shop* ($p < 0.05$) *Family income* ($p < 0.05$) *Use of eye kohl* (often) ($p < 0.01$) *Work in agriculture for more than 3 years* ($p < 0.01$)
Park et al. (2017), Korea	194	15 & 30 Dpp	[22–49]	Overall 15 Dpp (77) 30 Dpp (117)	8.79±14.1 13.3 ± 26 7.96 ± 13.2	4.71* (0.3–94.2) 4.82* (0.3–155) 4.11* (0.3–79)	Place of residence, lactation stage, maternal age, BMI Parity, smoking habits, and parturition	*Lactation stage* *Place of residence* ($p = 0.03$)

C colostrums; *T* transition milk; *M* mature milk; *Dpp* day postpartum; *Mpp* month postpartum; *Wpp* week postpartum; *W* women; ∑ Mean calculated from the total concentrations of the mothers; *ND* not detected; LC, MC, and HC indicate low, medium, and high consumption of local fish and fishery products; *PRT* premature; *FT* full term; *BP* before pregnancy

Table 2 Summary of the studies classified into affecting or not affecting factors and factors showing only differences of Pb level in human breast milk

Factors	Studies where factor is statistically significant ($p < 0.05$) or correlated*	Studies where the factor is showing a difference but not statistically significant	Studies where the factor did not affect the level of Pb	Number of total studies
Physiological factors				
Maternal age	Sternowsky and Wessolowski (1985), Sharma and Pervez (2005)*, Younes et al. (1995), Rahimi et al. (2009), Nassir et al. (2013), Winiarska-Mieczan (2014), Chao et al. (2014)*	Frković et al. (1997), Ettinger et al. (2004), (2005), Ursinyova and Masanova (2005)	Vanderjagt et al. (2001), Al-Saleh et al. (2003), Leotsinidis et al. (2005), Chien et al. (2006), Abdulrazzaq et al. (2008), Koyashiki et al. (2010), Orun et al. (2011), Garcia-Esquinas et al. (2011), Marques et al. (2013), Matos et al. (2014), Dursun et al. (2016), Shawahna et al. (2016), Park et al. (2017)	24
Lactation stage	Sternowsky and Wessolowski (1985), Krachler et al. (1998), Sowers et al. (2002), Ettinger et al. (2005)*, Leotsinidis et al. (2005), Winiarska-Mieczan (2014), Chao et al. (2014), Letinić et al. (2016), Park et al. (2017)	Ong et al. (1985), Younes et al. (1995), Friel et al. (1999), Chien et al. (2006), Kinsara et al. (2006), Abballe et al. (2008), Isaac et al. (2012)	Rockway et al. (1984), Rabinowitz et al. (1985), Namihira et al. (1993), Al-Saleh et al. (2003), Koyashiki et al. (2010), Marques et al. (2013), Abdulrazzaq et al. (2008), Shawahna et al. (2016)	24
Parity	Rahimi et al. (2009), Chao et al. (2014)*	Frković et al. (1997), Ettinger et al. (2004), Ursinyova and Masanova (2005)	Namihira et al. (1993), Vanderjagt et al. (2001), Al-Saleh et al. (2003), Ettinger et al. (2005), Leotsinidis et al. (2005), Chien et al. (2006), Orun et al. (2011), Garcia-Esquinas et al. (2011), Matos et al. (2014), Dursun et al. (2016), Park et al. (2017)	16

(continued)

Table 2 (continued)

Factors	Studies where factor is statistically significant ($p < 0.05$) or correlated*	Studies where the factor is showing a difference but not statistically significant	Studies where the factor did not affect the level of Pb	Number of total studies
Weight and height		Gundacker et al. (2002)	Al-Saleh et al. (2003), Leotsinidis et al. (2005), Chien et al. (2006), Abdulrazzaq et al. (2008), Orun et al. (2011)	6
BMI	Nassir et al. (2013)		Vanderjagt et al. (2001), Chien et al. (2006), Park et al. (2017)	4
Gestational age		Gundacker et al. (2002)	Friel et al. (1999), Orun et al. (2011), Garcia-Esquinas et al. (2011), Dursun et al. (2016)	5
Number of children	Namihira et al. (1993)*		Koyashiki et al. (2010), Marques et al. (2013), Shawahna et al. (2016)	4
Duration of previous breastfeeding	Al-Saleh et al. (2003)	Ettinger et al. (2004)	Ettinger et al. (2005), Garcia-Esquinas et al. (2011)	4
Parturition			Park et al. (2017), Leotsinidis et al. (2005)	2
Bone	Gulson et al. (1998b)			1
Skin color	Koyashiki et al. (2010)			1
Birth interval			Orun et al. (2011)	1
History of anemia	Orun et al. (2011)			1
External factors				

Place of residence	Huat et al. (1983), Hallén et al. (1995), Saleh et al. (1996), Frković et al. (1997), Al-Saleh et al. (2003), Gundacker et al. (2002), Leotsinidis et al. (2005), Sharma and Pervez (2005), Cinar et al. (2011), Isaac et al. (2012), Nassir et al. (2013), Marques et al. (2013), Shawahna et al. (2016), Park et al. (2017)	Sternowsky and Wessolowski (1985), Younes et al. (1995), Vavilis et al. (1997), Klopov (1997), Hamzaoglu et al. (2014)	Rockway et al. (1984), Turconi et al. (2004), Ursinyova and Masanova (2005), Abballe et al. (2008), Abdulrazzaq et al. (2008), Koyashiki et al. (2010), Orun et al. (2011), Garcia-Esquinas et al. (2011)	27
Years of residence in the area House age Residence near contaminant source	Marques et al. (2013)∗, Dabeka et al. (1986)∗, Shawahna et al. (2016)		Namihira et al. (1993), Ettinger et al. (2004), Abdulrazzaq et al. (2008), Koyashiki et al. (2010)	4 1 2
Maternal dietary intakes Coffee Milk Dairy products Cheese Rice Red meat Fish Canned food Vegetables Cereals Potato Beverages Traditional Chinese herbs Type of drinking water Wine Alcohol	Gulson et al. (1998b), Dabeka et al. (1986), Garcia-Esquinas et al. (2011), Leotsinidis et al. (2005), Al-Saleh et al. (2003), Gundacker et al. (2002), Garcia-Esquinas et al. (2011), Garcia-Esquinas et al. (2011), Chien et al. (2006), Nassir et al. (2013)	Gundacker et al. (2002)	Da silva Anastácio et al. (2004), Turconi et al. (2004), Abballe et al. (2008), Koyashiki et al. (2010), Dursun et al. (2016), Chao et al. (2014), Koka et al. (2011), Hallén et al. (1995), Chien et al. (2006), Shawahna et al. (2016)	17

(continued)

Table 2 (continued)

Factors	Studies where factor is statistically significant ($p < 0.05$) or correlated*	Studies where the factor is showing a difference but not statistically significant	Studies where the factor did not affect the level of Pb	Number of total studies
Kind of utensil copper casserole coated with white (rich in lead) inner coat and eating food preserved for long periods in metal containers Lead-glazed ceramics Lead-glazed ceramics and lead in home pumping Clay utensils	Ettinger (2004, 2005)	Younes et al. (1995)	Dursun et al. (2016), Shawahna et al. (2016)	5
Smoking habits	Gundacker et al. (2002), Koka et al. (2011), Nassir et al. (2013), Winiarska-Mieczan (2014), Letinić et al. (2016)	Ursinyova and Masanova (2005)	Frković et al. (1997), Hallén et al. (1995), Turconi et al. (2004), Chien et al. (2006), Abdulrazzaq et al. (2008), Koyashiki et al. (2010), Orun et al. (2011), Garcia-Esquinas et al. (2011), Chao et al. (2014), Rahimi et al. (2009), Dursun et al. (2016), Shawahna et al. (2016), Park et al. (2017)	19
Vitamins and minerals supplementation	da Silva Anastácio et al. (2004)	Gundacker et al. (2002), Ettinger et al. (2004)	Ettinger et al. (2005), Matos et al. (2014), Leotsinidis et al. (2005), Orun et al. (2011)	7
Dental amalgam fillings	Chao et al. (2014)		Ursinyova and Masanova (2005), Dursun et al. (2016), Garcia-Esquinas et al. (2011)	4
Exposition to heavy traffic	Dabeka et al. (1986), Garcia-Esquinas et al. (2011)			2
Exposition to insecticide	Dursun et al. (2016)		Chao et al. (2014)	2
Use of kohl	Shawahna et al. (2016)			1

Socioeconomic factors				
Maternal education level	Leotsinidis et al. (2005), Chao et al. (2014)*		Al-Saleh et al. (2003), Abdulrazzaq et al. (2008), Garcia-Esquinas et al. (2011), Orun et al. (2011), Shawahna et al. (2016)	7
Maternal work status			Al-Saleh et al. (2003), Abdulrazzaq et al. (2008), Orun et al. (2011), Dursun et al. (2016), Leotsinidis et al. (2005)	5
Kind of job/working in agriculture	Shawahna et al. (2016)		Turconi et al. (2004)	2
Family monthly income	Shawahna et al. (2016)		Al-Saleh et al. (2003), Koyashiki et al. (2010), Orun et al. (2011)	4
Newborn factors				
Gender			Ursinyova and Masanova (2005), Abdulrazzaq et al. (2008), Orun et al. (2011), Dursun et al. (2016)	4
Birth and current weight, length, and head circumference			Ursinyova and Masanova (2005), Al-Saleh et al. (2003), Leotsinidis et al. (2005), Chien et al. (2006), Dursun et al. (2016)	5

* means correlated

3.4 Factors Associated with the Level of Lead in Human Breast Milk

Literature findings concerning the different factors affecting lead concentrations in breast milk are very controversial (Table 2).

3.4.1 Place of Residence

The area of residence seemed to influence the level of Pb in HBM, and this factor was the most investigated in the reviewed studies (27 studies). The majority of the studies found that women living in urban or industrial areas had BM lead concentrations significantly higher than women leaving in rural or nonindustrial ones, except for three studies from Saudi Arabia (Al-Saleh et al. 2003) and Turkey (Cinar et al. 2011; Hamzaoglu et al. 2014). Statistically significant differences were reported in 14 studies from Malaysia, Sweden, Egypt, Croatia, India, Turkey, Iraq, Brazil, Palestine, and Korea (Al-Saleh et al. 2003; Cinar et al. 2011; Frković et al. 1997; Gundacker et al. 2002; Hallén et al. 1995; Huat et al. 1983; Isaac et al. 2012; Leotsinidis et al. 2005; Marques et al. 2013; Nassir et al. 2013; Park et al. 2017; Saleh et al. 1996; Sharma and Pervez 2005; Shawahna et al. 2016). Only five surveys displayed differences (Hamzaoglu et al. 2014; Klopov 1997; Sternowsky and Wessolowski 1985; Vavilis et al. 1997; Younes et al. 1995). However, other studies reported that the place of residence did not affect the level of Pb in HBM (Abballe et al. 2008; Abdulrazzaq et al. 2008; Garcia-Esquinas et al. 2011; Koyashiki et al. 2010; Orun et al. 2011; Rockway et al. 1984; Turconi et al. 2004; Ursinyova and Masanova 2005).

The majority of the findings mentioned above coincide with a study of the WHO (1989) which reported that the industrialized countries (Sweden, Hungary) had higher Pb concentrations in HBM than nonindustrialized ones (Parr et al. 1991). The Pb content of the mother's milk in strongly polluted areas can be 20 times higher (Sonawane 1995).

Indeed, Sternowsky and Wessolowski (1985) reported that urban women have mean levels that seemed to increase with duration of lactation, while in the rural area, they decreased (Sternowsky and Wessolowski 1985). Isaac et al. (2012) noticed a significant decline in the Pb levels in the breast milk of mothers of the nonindustrialized areas in comparison with samples from mothers of industrialized ones. Also, the lactating mothers from industrialized areas have lower IQ levels compared to mothers from nonindustrial ones (Isaac et al. 2012).

The length of residence in the area (Marques et al. 2013), the dwelling place near sources of contamination (Shawahna et al. 2016), as well the age of the house (Dabeka et al. 1986) showed significant correlation with Pb in HBM.

3.4.2 Maternal Age

The factor of maternal age was studied by many authors (25 studies) and significantly correlated and affected the level of Pb in HBM in seven studies. Also, the difference was observed in five papers, but the remaining studies did not find any association.

Furthermore, most of the studies reported that Pb levels were higher in older mothers (Chao et al. 2014; Ettinger et al. 2004, 2005; Rahimi et al. 2009; Sharma and Pervez 2005; Sternowsky and Wessolowski 1985; Winiarska-Mieczan 2014; Younes et al. 1995). However, younger mothers had the highest amount of Pb in BM in three observed surveys (Frković et al. 1997; Nassir et al. 2013; Ursinyova and Masanova 2005).

Indeed, older women accumulated higher amounts of lead due to a higher prolonged duration of exposure (ATSDR 1993). Women older than 30 years of age had significantly higher Pb levels in their BM than women between 20 and 30 years of age (Bonithon-Kopp et al. 1986). Such a difference may reflect the general increase in stored and circulating Pb levels as a function of age or altered mineral metabolism during lactation in older women (Silbergeld 1991).

3.4.3 Lactation Stage

Several studies assessed the influence of the lactation stage (24 studies). Furthermore, the majority reported that the Pb concentration decreased during the course of lactation and that the Pb level was higher in colostrum than transitory milk and mature milk (Chao et al. 2014; Ettinger et al. 2005; Krachler et al. 1998; Leotsinidis et al. 2005; Letinić et al. 2016; Matos et al. 2014; Park et al. 2017; Sowers et al. 2002; Sternowsky and Wessolowski 1985). The probable reason was due to the higher protein content of colostrum and the more significant amount of milk produced later in the lactation period (Rothenberg et al. 2000). However, the difference was statistically significant in nine studies. Seven studies showed minor differences (Almeida et al. 2008; Chien et al. 2006; Friel et al. 1999; Isaac et al. 2012; Kinsara et al. 2006; Ong et al. 1985; Younes et al. 1995). On the other hand, eight further surveys did not find any association, including two old studies from the United States (Rabinowitz et al. 1985; Rockway et al. 1984).

Indeed, a Saudi study mentioned that the Pb concentrations in milk samples collected in the afternoon were higher than those from the morning (Kinsara et al. 2006).

3.4.4 Smoking Habits

Gundacker et al. (2002), Rahimi et al. (2009), Koka et al. (2011), Nassir et al. (2013), Winiarska-Mieczan (2014), and Letinić et al. (2016) reported that the smoking habit of the mothers (past, current, or passive) significantly affected the level of

Pb in BM. However, the difference was not significant in the survey of Ursinyova and Masanova (2005), conflicting with the findings of 13 studies which did not establish any effect (Koka et al. 2011; Letinić et al. 2016; Nassir et al. 2013; Rahimi et al. 2009; Ursinyova and Masanova 2005; Winiarska-Mieczan 2014).

Letinić et al. (2016) reported that the lactation stage correlated to smoking habits and that the concentrations of Pb were almost twofold higher in colostrum of non-smokers and about 2.5 times higher in the transitional milk of smokers who continued smoking during pregnancy.

3.4.5 Dietary Habits

About 16 studies assessed the influence of maternal nutritional intakes on the level of Pb in HBM. Leotsinidis et al. (2005) found that consumption of red meat seems to decrease Pb level in BM and to be a protective factor and also that cheese and rice seem to influence positively the levels of Pb in BM and to increase it. Further, Garcia et al. (2011) observed an inverse association with ingestion of caffeinated coffee beverages and the level of Pb in BM. Accordingly, coffee, dairy products, cheese, rice, red meat, fish, potato, beverages, and traditional Chinese herbs were the foods that affected significantly (Al-Saleh et al. 2003; Chien et al. 2006; Dabeka et al. 1986; Garcia-Esquinas et al. 2011; Gundacker et al. 2002; Leotsinidis et al. 2005). Also, mothers who consumed cereals frequently had slightly higher Pb in their milk than mothers who occasionally or never consumed cereals, but the difference was not significant (Gundacker et al. 2002). Others indicated that fish (Abballe et al. 2008; Dursun et al. 2016), milk (Koyashiki et al. 2010), and dietary habits (da Silva Anastácio et al. 2004; Turconi et al. 2004) did not affect the level of Pb in BM.

Indeed, a single study found that tap and river water contributed to the increase of Pb in BM (Nassir et al. 2013). Nevertheless, another did not establish any association (Koka et al. 2011).

3.4.6 Parity

Many studies checked the relationship between parity and the level of Pb in HBM (16 studies). In fact, five studies demonstrated that multiparous females had higher Pb in their BM than primiparous ones, but the difference was significant only in two studies (Chao et al. 2014; Ettinger et al. 2004; Frković et al. 1997; Rahimi et al. 2009). Conversely, Ursinyova and Masanova (2005) reported that primiparous females had the highest levels of Pb in their milk. Indeed, ten studies mentioned that the parity did not change the Pb concentration.

A study from Mexico found a correlation between the number of children and the level of Pb in BM (Namihira et al. 1993), though three other studies failed to find any association.

3.4.7 Other Factors

Socioeconomic factors were explored by some researchers (Abdulrazzaq et al. 2008; Al-Saleh et al. 2003; Dursun et al. 2016; Garcia-Esquinas et al. 2011; Koyashiki et al. 2010; Leotsinidis et al. 2005; Orun et al. 2011; Shawahna et al. 2016; Turconi et al. 2004). However, only two studies from Greece and Taiwan declared that the level of maternal education influenced the level of Pb in BM (Chao et al. 2014; Leotsinidis et al. 2005), and a study from Palestine recorded that working in agriculture and the family income influenced too (Shawahna et al. 2016).

Regarding the maternal anthropometry, six studies concluded that weight and height did not affect the concentration of Pb in BM. On the other hand, Nassir et al. (2013) observed that overweight and obese mothers with a high body mass index (BMI) had higher Pb levels in their milk than mothers with average body weight. However, three other studies did not observe any differences (Chien et al. 2006; Park et al. 2017; VanderJagt et al. 2001). Furthermore, seven surveys declared that no association exists between factors in newborns (gender, weight, length, and head circumference) and Pb levels in HBM.

Indeed, Da Silva Anastácio et al. (2004) found a positive correlation with Ca intake, and Ettinger et al. (2004) reported that it affects the level of Pb in HBM. Moreover, Orun et al. (2011) mentioned that mothers with a history of anemia had higher Pb levels in their BM.

Two studies revealed important findings. The first one showed that the primary source of Pb in HBM is in the maternal bone, while the second one pointed out that there is a statistically significant association between the lead level in milk and the variable skin color and that white mothers had lower Pb in their milk than nonwhite mothers (Gulson et al. 1998a; Koyashiki et al. 2010).

The use of ceramics, clay, and metal utensils increased the level of Pb in HBM, but the difference was significant only in the studies by Ettinger et al. (2004, 2005).

Finally, exposure to heavy traffic (Dabeka et al. 1986; Garcia-Esquinas et al. 2011) and to insecticides (Dursun et al. 2016) and the use of kohl eyeliner often (Shawahna et al. 2016) significantly increased the level of Pb in HBM.

4 Conclusion

The present review identified several important factors that were controversial between authors across countries, and the analysis of these factors showed that there are few consistent findings that are supported by the strength of the significant difference between the studies (number of samples, the range of mother's age, different habits, etc.).

The most frequently assessed factors among studies and that are supported by the principal factors affecting lead level in BM were the following: (1) place of residence (higher in industrial and urban areas), (2) maternal age (higher in older mothers),

(3) stage of lactation (higher in colostrum), (4) smoking habits, (5) maternal dietary intakes, and (6) parity. Other factors were not enough studied and were considered minor because few surveys evaluated their impacts.

5 Summary

A systematic literature search in PubMed, Science Direct, and Google Scholar databases was performed to identify relevant studies, published in English, which investigated and explored factors affecting the level of lead in human milk among lactating women around the world. Forty-nine papers were rated and explored the effect of one or several factors on the level of lead in human milk from 28 countries between 1983 and 2017 and through Europe, Asia, America, and Africa, reviewing more than 5,000 subjects. Place of residence, maternal age, stage of lactation, smoking habits, maternal dietary intakes, and parity were the mostly assessed factors among the studies and considered as the main factors affecting Pb levels in BM. However, the literature findings are very controversial.

References

Abballe A, Ballard TJ, Dellatte E, Di Domenico A, Ferri F, Fulgenzi AR et al (2008) Persistent environmental contaminants in human milk: concentrations and time trends in Italy. Chemosphere 73:S220–S227

Abdulrazzaq YM, Osman N, Nagelkerke N, Kosanovic M, Adem A (2008) Trace element composition of plasma and breast milk of well-nourished women. J Environ Sci Health A 43 (3):329–334

Almeida AA, Lopes CM, Silva AM, Barrado E (2008) Trace elements in human milk: correlation with blood levels, inter-element correlations and changes in concentration during the first month of lactation. J Trace Elem Med Biol 22:196–205

Al-Saleh I, Shinwari N, Mashhour A (2003) Heavy metal concentrations in the breast milk of saudi women. Biol Trace Elem Res 96:21–37

Anderson HA, Wolff MS (2000) Environmental contaminants in human milk. J Expo Sci Environ Epidemiol 10:755

ATSDR (1993) Toxicological profile for lead. Agency for toxic substances and disease registry. Division of Toxicology. Toxicology Information Branch, Atlanta

Bonithon-Kopp C, Huel G, Grasmick C, Sarmini H, Moreau T (1986) Effects of pregnancy on the inter-individual variations in blood levels of lead, cadmium and mercury. Biol Res Pregnancy Perinatol 7:37–42

Byczkowski JZ, Gearhart JM, Fisher JW (1994) "Occupational" exposure of infants to toxic chemicals via breast milk. Nutrition 10:43–48

Chao H, Guo CH, Huang CB, Chen PC, Li HC, Hsiung DY et al (2014) Arsenic, cadmium, lead, and aluminium concentrations in human milk at early stages of lactation. Pediatr Neonatol 55:127–134

Chien L-C, Yeh C-Y, Lee H-C, Chao HJ, Shieh M-J, Han B-C (2006) Effect of the mother's consumption of traditional chinese herbs on estimated infant daily intake of lead from breast milk. Sci Total Environ 354(2–3):120–126

Cinar N, Ozdemir S, Yucel O, Ucar F (2011) In which regions is breast-feeding safer from the impact of toxic elements from the environment? Bosn J Basic Med Sci 11:234–239

da Silva Anastácio A, Da Silveira CLP, Miekeley N, Donangelo CM (2004) Distribution of lead in human milk fractions. Biol Trace Elem Res 102:27–37

Dabeka R, Karpinski K, McKenzie A, Bajdik C (1986) Survey of lead, cadmium and fluoride in human milk and correlation of levels with environmental and food factors. Food Chem Toxicol 24:913–921

Dillon HK, Wilson DJ, Schaffner W (1974) Lead concentrations in human milk. Am J Dis Child 128:491–492

Dursun A, Yurdakok K, Yalcin SS, Tekinalp G, Aykut O, Orhan G et al (2016) Maternal risk factors associated with lead, mercury and cadmium levels in umbilical cord blood, breast milk and newborn hair. J Matern Fetal Neonatal Med 29:954–961

Ettinger AS, Téllez-Rojo MM, Amarasiriwardena C, González-Cossío T, Peterson KE, Aro A et al (2004) Levels of lead in breast milk and their relation to maternal blood and bone lead levels at one month postpartum. Environ Health Perspect 112:926

Ettinger AS, Téllez-Rojo MM, Amarasiriwardena C, Peterson KE, Schwartz J, Aro A et al (2005) Influence of maternal bone lead burden and calcium intake on levels of lead in breast milk over the course of lactation. Am J Epidemiol 163:48–56

Friel JK, Andrews WL, Jackson SE, Longerich HP, Mercer C, McDonald A et al (1999) Elemental composition of human milk from mothers of premature and full-term infants during the first 3 months of lactation. Biol Trace Elem Res 67:225–247

Frković A, Kraš M, Alebić-Juretić A (1997) Lead and cadmium content in human milk from the northern adriatic area of croatia. Bull Environ Contam Toxicol 58:16–21

Garcia-Esquinas E, Perez-Gomez B, Fernandez MA, Perez-Meixeira AM, Gil E, de Paz C et al (2011) Mercury, lead and cadmium in human milk in relation to diet, lifestyle habits and sociodemographic variables in madrid (Spain). Chemosphere 85:268–276

Gulson BL, Jameson CW, Mahaffey KR, Mizon KJ, Patison N, Law AJ et al (1998a) Relationships of lead in breast milk to lead in blood, urine, and diet of the infant and mother. Environ Health Perspect 106:667

Gulson B, Mahaffey K, Jameson CW, Mizon K, Korsch M, Cameron M et al (1998b) Mobilization of lead from the skeleton during the postnatal period is larger than during pregnancy. J Lab Clin Med 131:324–329

Gundacker C, Pietschnig B, Wittmann KJ, Lischka A, Salzer H, Hohenauer L et al (2002) Lead and mercury in breast milk. Pediatrics 110:873–878

Hallén IP, Jorhem L, Lagerkvist BJ, Oskarsson A (1995) Lead and cadmium levels in human milk and blood. Sci Total Environ 166:149–155

Hamzaoglu O, Yavuz M, Turker G, Savli H (2014) Air pollution and heavy metal concentration in colostrum and meconium in two different districts of an industrial city: a preliminary report. Intern Med J:21

Huat LH, Zakariya D, Eng KH (1983) Lead concentrations in breast milk of malaysian urban and rural mothers. Arch Environ Health 38:205–209

IARC (2006) Iarc monographs on the evaluation of the carcinogenic risks of chemical to human. Inorganic and organic lead compounds. IARC, Lyon

Isaac CPJ, Sivakumar A, Kumar CRP (2012) Lead levels in breast milk, blood plasma and intelligence quotient: a health hazard for women and infants. Bull Environ Contam Toxicol 88:145–149

Järup L (2003) Hazards of heavy metal contamination. Br Med Bull 68:167–182

Kinsara A, Farid S, Wajid S, Sadi S (2006) Longitudinal variation of trace elements concentration in human milk in Jeddah, Saudi Arabia. Eng Sci 17:75–94

Klopov VP (1997) Levels of heavy metals in women residing in the russian arctic. Int J Circumpolar Health 57:582–585

Koka J, Koranteng-Addo J, Bentum J, Koka D, Kamoah G (2011) Analysis of lead and cadmium in human milk in the greater accra region of Ghana. Der Chemica Sinica 2(2):240–246

Koyashiki GAK, Paoliello MMB, Matsuo T, de Oliveira MMB, Mezzaroba L, de Fátima Carvalho M et al (2010) Lead levels in milk and blood from donors to the breast milk bank in Southern Brazil. Environ Res 110:265–271

Krachler M, Shi Li F, Rossipal E, Irgolic KJ (1998) Changes in the concentrations of trace elements in human milk during lactation. J Trace Elem Med Biol 12:159–176

Leotsinidis M, Alexopoulos A, Kostopoulou-Farri E (2005) Toxic and essential trace elements in human milk from greek lactating women: association with dietary habits and other factors. Chemosphere 61:238–247

Letinić JG, Sarić MM, Piasek M, Jurasović J, Varnai VM, Grgec AS et al (2016) Use of human milk in the assessment of toxic metal exposure and essential element status in breastfeeding women and their infants in coastal croatia. J Trace Elem Med Biol 38:117–125

Marcus AH (1985) Multicompartment kinetic models for lead: II. Linear kinetics and variable absorption in humans without excessive lead exposures. Environ Res 36:459–472

Marques RC, Maria de Fátima RM, Bernardi JVE, Dórea JG (2013) Breast milk lead concentrations of mothers living near tin smelters. Bull Environ Contam Toxicol 91:549–554

Matos C, Moutinho C, Almeida C, Guerra A, Balcao V (2014) Trace element compositional changes in human milk during the first four months of lactation. Int J Food Sci Nutr 65:547–551

Moher D, Liberati A, Tetzlaff J, Altman DG, PRISMA Group (2009) Preferred reporting items for systematic reviews and meta-analyses: the PRISMA statement. PLoS Med 6:e1000097

Namihira D, Saldivar L, Pustilnik N, Carreon G, Salinas M (1993) Lead in human blood and milk from nursing women living near a smelter in Mexico city. J Toxicol Environ Health 38:225–232

Nassir IM, Al-Sharify AN, Baiee HA (2013) Lead and cadmium in the breast milk of lactating mothers living in Hilla City, Babylon, Iraq, during the year 2012. J Univ Babylon 21(8):2861–2872

Ong C, Phoon W, Law H, Tye C, Lim H (1985) Concentrations of lead in maternal blood, cord blood, and breast milk. Arch Dis Child 60:756–759

Orun E, Yalcin SS, Aykut O, Orhan G, Morgil GK, Yurdakok K et al (2011) Breast milk lead and cadmium levels from suburban areas of Ankara. Sci Total Environ 409:2467–2472

Park Y, Lee A, Choi K, Kim H-J, Lee JJ, Choi G et al (2017) Exposure to lead and mercury through breastfeeding during the first month of life: a check cohort study. Sci Total Environ 612:876–883

Parr RM, DeMaeyer EM, Iyengar VG, Byrne AR, Kirkbright GF, Schöch G et al (1991) Minor and trace elements in human milk from Guatemala, Hungary, Nigeria, Philippines, Sweden, and Zaire. Biol Trace Elem Res 29:51–75

Rabinowitz M, Leviton A, Needleman H (1985) Lead in milk and infant blood: a dose-response model. Arch Environ Health 40:283–286

Rahimi E, Hashemi M, Baghbadorani ZT (2009) Determination of cadmium and lead in human milk. Int J Environ Sci Tech 6:671–676

Rockway SW, Weber CW, Lei KY, Kemberling SR (1984) Lead concentrations of milk, blood, and hair in lactating women. Int Arch Occup Environ Health 53:181–187

Rothenberg SJ, Khan F, Manalo M, Jiang J, Cuellar R, Reyes S et al (2000) Maternal bone lead contribution to blood lead during and after pregnancy. Environ Res 82:81–90

Saleh MA, Ragab AA, Kamel A, Jones J, El-Sebae AK (1996) Regional distribution of lead in human milk from egypt. Chemosphere 32:1859–1867

Sharma R, Pervez S (2005) Toxic metals status in human blood and breast milk samples in an integrated steel plant environment in central india. Environ Geochem Health 27:39–45

Shawahna R, Zyoud A, Dwikat J, El-Helo M, Yacoub B, Hilal H (2016) Breast milk lead levels in 3 major regions of the west bank of palestine. J Hum Lact 32:455–461

Silbergeld EK (1991) Lead in bone: implications for toxicology during pregnancy and lactation. Environ Health Perspect 91:63

Silbergeld EK, Schwartz J, Mahaffey K (1988) Lead and osteoporosis: mobilization of lead from bone in postmenopausal women. Environ Res 47:79–94

Sonawane BR (1995) Chemical contaminants in human milk: an overview. Environ Health Perspect 103:197

Sowers MR, Scholl TO, Hall G, Jannausch ML, Kemp FW, Li X et al (2002) Lead in breast milk and maternal bone turnover. Am J Obstet Gynecol 187:770–776

Sternowsky HJ, Wessolowski R (1985) Lead and cadmium in breast milk higher levels in urban vs rural mothers during the first 3 months of lactation. Arch Toxicol 57:41–45

Turconi G, Guarcello M, Livieri C, Comizzoli S, Maccarini L, Castellazzi AM et al (2004) Evaluation of xenobiotics in human milk and ingestion by the newborn – an epidemiological survey in lombardy (northern italy). Eur J Nutr 43:191–197

Ursinyova M, Masanova V (2005) Cadmium, lead and mercury in human milk from Slovakia. Food Addit Contam 22:579–589

VanderJagt DJ, Okolo SN, Romero L, Millson M, Glew RH (2001) Lead levels in the milk of fulani women in Nigeria. J Natl Med Assoc 93:104

Vavilis D, Bontis J, Agorastos T, Angelikakis G, Zournatzi V, Loufopoulos A et al (1997) Lead concentrations in early human milk of urban and rural mothers. Clin Exp Obstet Gynecol 24:198–199

Winiarska-Mieczan A (2014) Cadmium, lead, copper and zinc in breast milk in poland. Biol Trace Elem Res 157:36–44

World Health Organization (WHO) (1989) Minor and trace elements in breast milk: report of a joint WHO/IAEA collaborative study. World Health Organization, Geneva

World Health Organization (WHO) (1996) Trace elements in human nutrition and health. World Health Organization, Geneva

Younes B, Al-Meshari AA, Al-Hakeem A, Al-Saleh S, Al-Zamel F, Al-Shammari F et al (1995) Lead concentration in breast milk of nursing mothers living in Riyadh. Ann Saudi Med 15 (3):249–251

Current Status and Future Prognosis of Malaria Vector Control Pesticide Ecotoxicology and *Xenopus* sp.

Nico J. Wolmarans, Lieven Bervoets, Patrick Meire, and Victor Wepener

Contents

N. J. Wolmarans (✉)
Water Research Group, Unit for Environmental Sciences and Management, North-West University, Potchefstroom, South Africa

Laboratory of Systemic, Physiological and Ecotoxicological Research, Department of Biology, University of Antwerp, Antwerp, Belgium
e-mail: 21600600@nwu.ac.za; NicolaasJohannes.Wolmarans@uantwerpen.be

L. Bervoets
Laboratory of Systemic, Physiological and Ecotoxicological Research, Department of Biology, University of Antwerp, Antwerp, Belgium
e-mail: lieven.bervoets@uantwerpen.be

© Springer Nature Switzerland AG 2019

P. de Voogt (ed.), *Reviews of Environmental Contamination and Toxicology, Volume 252*, Reviews of Environmental Contamination and Toxicology Volume 252,
https://doi.org/10.1007/398_2019_35

Abbreviations

ACP	Acid phosphatase
AI	Active ingredient
AMA	Amphibian metamorphosis assay
CaE	Carboxylesterase
CSA	Cockayne syndrome A gene
DDA	Dichlorodiphenylacetic acid
DDD	Dichlorodiphenyldichloroethane
DDE	Dichlorodiphenyldichloroethylene
DDT	Dichlorodiphenyltrichloroethane
DNA	Deoxyribonucleic acid
EC50	Effective concentration where 50% of the test population are affected
EDC	Endocrine-disrupting compound
FETAX	Frog embryo teratogenesis assay-*Xenopus*
GST	Glutathione-S-transferase
IC50	Inhibition concentration where 50% of the test population show inhibition of a measured aspect
IRS	Indoor residual spraying
ITN	Insecticide-treated net
LAGDA	Larval amphibian growth and development assay
LC50	Lethal concentration where 50% of the test population died
LDH	Lactate dehydrogenase
LOEC	Lowest observed effects concentration
MCIG	Minimum concentration to inhibit growth
m-RNA	Messenger ribonucleic acid
MUTL	Muir-Torre syndrome gene
MVC	Malaria vector control
MVCP	Malaria vector control pesticide
NAD^+	Nicotinamide adenine dinucleotide
NF	Nieuwkoop-Faber
OECD	Organisation for Economic Co-operation and Development
POP	Persistent organic pollutant
SSD	Species sensitivity distribution
TI	Teratogenic index
UVB	Ultraviolet B

P. Meire
Ecosystem Management Research Group (Ecobe), Department of Biology, University of Antwerp, Antwerp, Belgium
e-mail: partick.meire@uantwerpen.be

V. Wepener
Water Research Group, Unit for Environmental Sciences and Management, North-West University, Potchefstroom, South Africa
e-mail: Victor.wepener@nwu.ac.za

WHO	World Health Organization
XPA	Xeroderma pigmentosum group A gene
XPG	Xeroderma pigmentosum group G gene

1 Problem Statement

While frogs are generally considered as good indicators of environmental change, they are not widely utilised as indicators in ecological studies. With much of current research relying on model organisms as proxies for other animals or even humans, toxicology data on these model organisms is important for higher tier risk assessment studies and ecological modelling. The issue at hand is that for some topics, even when looking at model organisms, this empirical data does not seem to be as common as expected when doing literature research. Malaria vector control (MVC) is often a sensitive subject due to the paradox surrounding it (Bouwman et al. 2011). The lack of toxicological data mentioned feeds into the issues complicating debates on the chemical vector control interventions. Although being in use for decades, there are still many unknowns regarding the environmental impact of these pesticides. Informed decision-making and policies require that this information first be available. The main research problem investigated in this review is this lack of toxicological and biomonitoring data available for one of the model organisms (*Xenopus laevis*) representative of aquatic ecosystems, and to a lesser extent, anurans. Species from the *Xenopus* genus are endemic to Africa, and their distribution includes many areas where malaria vector control is applied. Reliable toxicology data on these frogs regarding malaria vector control pesticides (MVCPs) would strengthen the use of these animals as bio-indicators.

2 Introduction

The global fight against malaria, nature's most prolific killer, is largely structured around vector control (i.e. controlling mosquito populations). This entails the use of various methods including biological control through the use of larvivorous fish, the efficacy of which has not yet been validated (Walshe et al. 2017). But, for the most part, vector control globally relies on chemical control through the use of various insecticides. These MVCPs include persistent and non-persistent compounds from various pesticide classes. Pesticides are used individually or in combination and through different application methods. The most common and effective chemical methods of intervention are indoor residual spraying (IRS) and insecticide-treated nets (ITNs) (Glunt et al. 2013). The World Health Organization (WHO) Global Malaria Programme lists 12 recommended insecticides for vector control (Table 1). These insecticides belong to four chemical groups: organochlorines, pyrethroids, organophosphates and carbamates (WHO 2006). The use of these pesticides also

Table 1 List of insecticides recommended by the World Health Organization (WHO) for use for indoor residual spraying against malaria vectors

Insecticide	CAS#	Formulations	Chemical class	Dosage (g AI/m^2)	Mode of action	Effective duration (months)
α-Cypermethrin	67375-30-8	Wettable powder/suspension concentrate	Pyrethroid	0.02–0.03	Contact	4–6
λ-Cyhalothrin	91465-08-6	Wettable powder/capsule suspension	Pyrethroid	0.02–0.03	Contact	3–6
Bendiocarb	22781-23-3	Wettable powder	Carbamate	0.1–0.4	Contact/airborne	2–6
Bifenthrin	82657-04-3	Wettable powder	Pyrethroid	0.025–0.05	Contact	3–6
Cyfluthrin	68359-37-5	Wettable powder	Pyrethroid	0.02–0.05	Contact	3–6
DDT	8017-34-3	Wettable powder	Organochlorine	1–2	Contact	>6
Deltamethrin	52918-63-5	Wettable powder/water dispersible granule	Pyrethroid	0.02–0.025	Contact	3–6
Etofenprox	80844-07-1	Wettable powder	Pyrethroid	0.1–0.3	Contact	3–6
Fenitrothion	122-14-5	Wettable powder	Organophosphate	2	Contact/airborne	3–6
Malathion	121-75-5	Wettable powder	Organophosphate	2	Contact	2–3
Pirimiphos-methyl	29232-93-7	Wettable powder/emulsifiable concentrate	Organophosphate	1–2	Contact/airborne	2–3
Propoxur	114-26-1	Wettable powder	Carbamate	1–2	Contact/airborne	3–6

Adapted from WHO original list: http://www.who.int/whopes/Insecticides_IRS_Malaria_09.pdf

varies between countries based on policy. Recently there has been a global decline in the use of pesticides for IRS, partly as efforts to reduce selection resistance in mosquitoes (Alonso and Noor 2017). Efforts to find alternative insecticides started in the late 1960s to early 1970s. Driven mainly by resistance in mosquitoes and subsequently by the secondary environmental effects of some pesticides such as DDT, the first official effort by the WHO to explore alternative insecticides was at the International Conference on Alternative Insecticides for Vector Control in 1971 (WHO 1971). Here it was stated that ideal insecticides should be of low toxicity to mammals and high toxicity to insects; however, not much focus was drawn to the effects on lower vertebrates. Acetylcholinesterase acting insecticides such as carbamates and organophosphates received major attention as the first alternatives, along with pyrethroids. The development of new synthetic pyrethroids introduced this group as a promising alternative. The pyrethroids currently recommended by the WHO for malaria vector control are considered the more advanced synthetic pyrethroids able to show better photostability and potency towards mosquitoes than the natural pyrethroids (pyrethrin I and II; CAS: 121-21-1 and CAS: 121-29-9). The chirality of pyrethroids plays a crucial role in their activity and stability. The ability to separate single isomers allowed for more targeted efficacy such as with bioresmethrin (CAS: 28434-01-7), the 1R-*trans* isomer of resmethrin (CAS: 10453-86-8), showing very low toxicity to mammals, yet equal efficacy against insects to the parent resmethrin (Knaak et al. 2012). The addition of the a-cyano group in the development of cypermethrin (CAS: 67375-30-8) and deltamethrin (CAS: 52918-63-5) greatly increased photostability of these insecticides (Knaak et al. 2012).

Regardless of efforts made to improve on vector control insecticides since the 1970s, a "golden bullet" that is highly toxic to mosquitoes, is persistent in its effect on insects, has no resistance build-up in mosquito populations and is with zero to minimal non-target effects has not been discovered. The nature of pesticides and some application processes used, such as spraying a water-based solution, causes relatively easy transfer to surrounding environments, especially aquatic environments that tend to act as sinks for accumulation of chemical pollution (Van der Oost et al. 2003). Lambert (2001) reported on amphibian deaths in Africa with regard to pesticide usage. Although pesticides may not be the only factors involved, some of the reported deaths occurred even after lower application rates than those recommended for MVC by the WHO (Table 1). The principle of introducing neurotoxic chemical agents to any ecosystem (aquatic or terrestrial) on a regular basis inevitably results in a potential risk to non-target organisms in that ecosystem. To assess these potential effects, model organisms are often used. This review discusses anurans from the genus *Xenopus*, with particular focus on *Xenopus laevis* (African clawed frog) and its use as a model organism in research regarding MVCPs around the world.

2.1 *History of* Xenopus *Use in Research*

Xenopus laevis, the frog species most commonly used in research, is endemic to southern Africa. Starting in the late 1930s, *X. laevis* was distributed to laboratories worldwide, since they served as the most reliable pregnancy test until the 1960s (Gurdon and Hopwood 2000). With laboratory housing conditions becoming refined, and *X. laevis* being distributed to laboratories around the world, it quickly became a model for scientific teaching and research. Its use in pregnancy testing relied on ovulation being induced in the female frogs, which paved the way for harvesting and artificial insemination of the ova for regulated breeding. Nieuwkoop and Faber (1994) studied the development of *X. laevis* in great detail. They produced a normal development table for 66 stages of *X. laevis* development from fertilisation to adulthood, first published in 1956, with a final refined version in 1994. This opened the door for the use of *X. laevis* as a model for frog development studies.

The immune system of *Xenopus* changes drastically as it develops (Robert and Ohta 2009), which may cause sensitivity to the teratogenic effects of compounds they are exposed to or even increase susceptibility to secondary infection (Mann et al. 2009). *Xenopus laevis* tadpoles are filter feeders, which increase their uptake of chemicals in the water column. Dumont et al. (1983) used the Nieuwkoop-Faber developmental staging model and developed a standardised test called the frog embryo teratogenesis assay-*Xenopus* (FETAX) (updated document: ASTM 2012), which allows for the measurement of developmental toxicity at different degrees through mortality, morphological changes and growth rate. All the measured morphological changes, and the severity thereof, were categorised by Hu et al. (2015). Most of the available data produced on amphibian ecotoxicology regarding this species is based on FETAX.

The validity of *X. laevis* as a model frog can be questioned, besides its use in FETAX. Two arguments against its use as a model are, firstly, that it is a fully aquatic species and therefore is not necessarily representative of anurans as a whole. Secondly, adult *X. laevis* are known as being hardy and can tolerate a wider range of physicochemical water parameters and, in some cases, toxicants as well, compared to other anurans. The basis of FETAX, however, is the sensitivity of the embryos and tadpoles. This indicates that the life stage examined should be considered carefully when *X. laevis* is used as test model. The context in which a study is done can also validate the use of the species, as members of the *Xenopus* genus can be seen as good representatives of freshwater organisms in general. This is based on the mid-level trophic position they hold in aquatic ecosystems (Lindholm et al. 2007; Wolmarans et al. 2015; Dalu et al. 2016). Their permeable skin can lead to increased dermal uptake, which also makes them good models for bioaccumulation studies. Increased uptake plays a large role when investigating toxicants with high affinity to sediments. These frogs tend to dwell in the sediment with their highly permeable belly skin exposed for extended periods. They also eat by stuffing food into their mouths and ingesting it by means of a hyobranchial pump (Avila and Frye 1978), increasing the likelihood of sediment ingestion and subsequent chemical exposure. *Xenopus* laevis tadpoles are filter feeders, which also increases sediment ingestion throughout their development. Dalu et al. (2016) showed, through means of stable isotope ratio analysis, that sediments along with coarse particulate organic matter contribute quite substantially to the diet of *X. laevis* tadpoles in ephemeral ponds, with sediment ranging between 1 and 33% and coarse particulate organic matter up to 40.1% of their diet.

The relative ease of housing, artificially harvesting and fertilising eggs and reproducing these organisms in a laboratory setting, as described by ASTM (2012), has also played a role in the continued use of this species. The introduction of *X. tropicalis* (*Silurana tropicalis* in some literature) as a replacement for *X. laevis* in some research studies in the late 1990s came about largely due the tetraploidy of *X. laevis*, which hindered research in genetic and gene expression fields (Burggren and Warburton 2007). *Xenopus tropicalis* is a diploid and has a faster development timeframe ($\approx$6 months to adulthood vs $\approx$ 12 months for *X. laevis*) and can also be used effectively with FETAX (Hu et al. 2015). Regardless of the use of *Xenopus* sp. in other types of studies, the FETAX model is still considered one of the primary assays to be used for testing teratogenic effects of compounds on lower vertebrate animals. Bearing this information in mind, it is interesting to note that the number of publications produced using *Xenopus* spp. as a model for pesticide research is quite low (Table 2). Although Boolean search systems (search system use is explained under availability of literature) for locating electronically available publications in online databases do not necessarily provide an accurate number of publications available on a subject, it gives a good indication of whether the topic is popularly explored by scientists around the world.

2.2 *Scope of the Review*

In this review we summarise the literature available on *Xenopus* sp. in relation to the 12 WHO-recommended MVCPs in terms of acute, sub-lethal and chronic toxicity. As supplementary information (Figs. 1 and 2), these data were evaluated using meta-analyses for theoretical relative toxicity of the different pesticides to *Xenopus* sp. in order to pinpoint which pesticides might require more focus for future research. This review serves to bring to attention the lack of available literature in this field and the inconsistency of results between some of the existing records. We also identify gaps in the current knowledge, in order to promote more focused future research in this field. Zippel and Mendelson (2008) proposed certain challenges to the global amphibian crisis, one of which was directly aimed at environmental contamination in generating data on sub-lethal and chronic effects. Now, a decade later, we hope that this review can stress the lack of progress made in this regard in the context of the pesticides in question.

2.3 *Data Inclusion Methods and Criteria*

Search queries were performed on Scopus, Web of Science and other online databases using selected keywords (Scopus search outcomes given in Table 2) pertaining to MCVPs and *Xenopus*. The US-EPA ECOTOX database was also queried, with Pauli et al. (2000) as secondary sources. Some criteria were necessary in the filtering of data used for this review. To be acceptable for inclusion, a study had to report on *Xenopus* sp. in relation to one or more of the 12 WHO-recommended malaria vector control pesticides listed in Table 1. Studies reporting toxicological effects without any corresponding reference as to

Table 2 Scopus database search outputs using the search field "Article, Abstract and Keywords" for various search phrases pertaining to *Xenopus* spp. and malaria vector control pesticides

Search phrase	Year	Records found	Records within review criteria
Xenopus AND cypermethrin	1982–2018	22	8
Xenopus AND cyhalothrin	2006–2012	5	1
Xenopus AND bendiocarb	2003	1	0
Xenopus AND bifentrhin	2006–2017	3	0
Xenopus AND cyfluthrin	1986–2017	7	2
Xenopus AND DDT	1969–2017	34	16
Xenopus and deltamethrin	1979–2017	47	8
Xenopus and etofenprox	2010	1	0
Xenopus and fenitrothion	1981	1	1
Xenopus AND malathion	1989–2016	11 (+ 1 duplicate record)	10
Xenopus AND propoxur	1994 and 2003	2	0
Xenopus AND DDT AND bioaccumulation	1995 and 2006	2	2
Xenopus AND deltamethrin AND bioaccumulation	2007	1	1
Xenopus AND DDT AND field	2004 and 2009	2	1
Xenopus AND deltamethrin AND field	2014	1	0
Xenopus AND malathion AND field	1994	1	0
Xenopus AND propoxur AND field	1994	1	0

Values were obtained from searches performed on 13 August 2018. Every combination of Xenopus and each of the 12 pesticides in question along with the optional third search words were queried. Search phrase combinations providing zero results were not included in this table

exposure amount or biological load were excluded. Laboratory and field-generated data were included. As dichlorodiphenyltrichloroethane (DDT; CAS: 8017-34-3), one of the pesticides included, is considered a persistent organic pollutant (POP), studies reporting only bioaccumulation data were also included. The main breakdown products of DDT, dichlorodiphenyldichloroethane (DDD; CAS: 72-54-8) and dichlorodiphenyldichloroethylene (DDE; CAS:72-55-9), can be even more persistent in the environment and are therefore indirectly introduced via DDT introduction. Studies reporting sub-lethal effects of these compounds were also included in the study. The majority of studies within search parameters that were excluded from this review involved transgenic *Xenopus* oocytes used merely as a medium in which

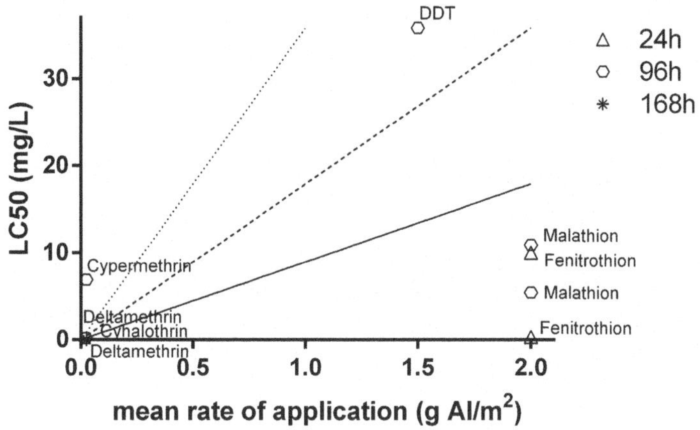

Fig. 1 Scatterplot of acute toxicity to *Xenopus laevis* vs recommended application rates for malaria vector control pesticides with relative potency categories indicated. Top quadrant shows lowest potency; bottom quadrant shows greatest potency relative to the other data on the graph. ··· = very low to low potency, --- = low to moderate potency and — = moderate to high potency

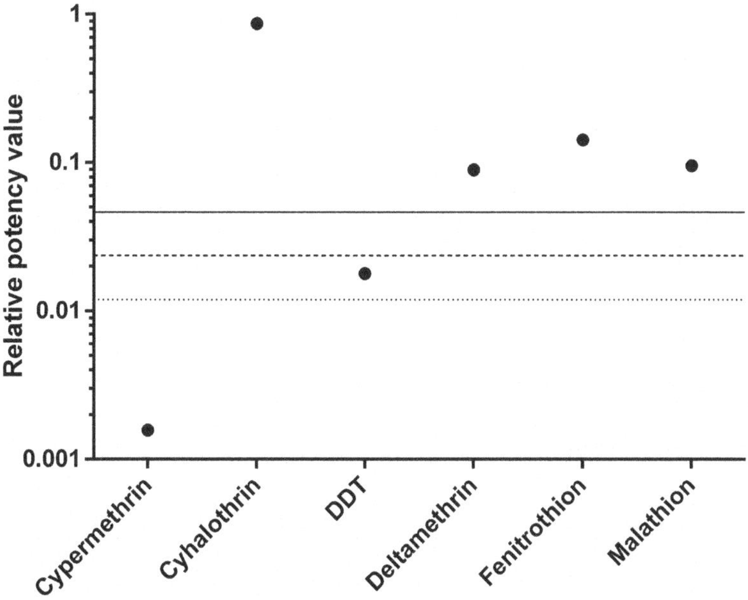

Fig. 2 Relative potency values for malaria vector control pesticides for causing acute toxic effects to *Xenopus laevis* calculated as Log((mean application rate × 1/mean LC50) + 1) with relative potency categories indicated as in Fig. 1. ··· = very low to low potency, --- = low to moderate potency and — = moderate to high potency

to analyse the reaction of transplanted genes to pesticides. If such studies did not report effects relatable to *Xenopus* sp. itself, they were excluded from this review.

2.4 Availability of Literature

Amphibian ecotoxicology is a research topic that has increased in popularity over the last decade, a trend also seen in literature included in this review. A Scopus database query performed in August 2018 using the search phrase "anura AND pesticide" in the field "Abstract, title and keywords" returned 395 entries. When the more refined search term "*Xenopus* AND insecticide" was used, 267 entries were given. Replacing "insecticide" with the names of the 12 WHO-recommended MVCPs, the highest output was for "*Xenopus* AND deltamethrin" with 47 entries. While this number may seem high, only eight of these publications fitted the criteria for inclusion in this review (see search results provided in Table 2). Through individual pesticide and *Xenopus* combination searches, 135 entries were given in total, of which only 46 entries fitted the criteria for inclusion in this review.

Globally, amphibians are widely distributed, with their highest species density being found in tropical and sub-tropical areas (FEOW 2008). These are also the regions with temperatures suitable for transmission of *Plasmodium* sp. throughout the world (MAP 2018), yet there seems to be a lack of literature on amphibians – especially from tropical and sub-tropical regions – and on ecotoxicology with regard to MVCPs. With overlap in pesticide use between MVC, vector control for other tropical diseases and agriculture, the need for data from species relevant to these tropical and subtropical systems becomes clear. In the case of Africa, *Xenopus* sp. is a good starting point for collecting relevant anuran data.

3 Summary and Discussion of Reviewed Literature

3.1 FETAX Studies

With FETAX long being the most commonly used teratogenic toxicity model, it would be expected that most available literature on *Xenopus* and MVCP toxicity would originate from FETAX studies. Another advantage of this type of data is comparability, as FETAX allows for a standardised testing method with standardised endpoints. This is not really the case with MVCPs, though the literature available is much scarcer than expected, lowering the confidence of comparisons between data.

FETAX data relating to MVCPs available in literature are summarised in Table 3. Data relating to only 3 of the 12 WHO-recommended MVCPs could be found, namely, DDT, deltamethrin and malathion (CAS: 121-75-5).

Table 3 Summary of results from studies included in this review which performed teratogenic toxicity exposures on *Xenopus laevis* embryos with regard to any of the 12 WHO-recommended MVCPs using the FETAX protocol

Pesticide	Exposure range (mg/L)	LC50 (95% CI; mg/L)	EC50 (mg/L)	TI	MCIG	LOEC	Notes	Source
Deltamethrin	NR	0.19 (0.16–0.24)	0.006	31	NR	NR		Channing (1998)
Malathion	0.1–100	10.9 (10.6–11.3)	P = 0.33 G = 0.79 N = 2.16	P = 33.2 N = 5.1 G = 13.8	NR	NR	P = effect of abnormal pigmentation observed; G = effect of abnormal gut observed; N = abnormal notochord	Snawder and Chambers (1989)
Malathion	0.05–6.0	NR	2.39	NR	NR	NR	Mortality at 6 mg/L = 12%	Bonfanti et al. (2004)
p,p-DDT	0.39–99.9	35.8 (31.3–41.12)	14.7 (10.7–20.8)	2.4	1.5	6.24	LOEC = malformation	Saka (2004)
p,p-DDD	0.486–31.2	14.11 (12.0–16.89)	4.7 (3.3–6.9)	3	1.95	1.95	LOEC = malformation	Saka (2004)
p,p-DDE	1.24×10^{-4}–124.9	>124.98	>124.98	NR	1.25–12.5	NR		Saka (2004)

NR not reported

3.1.1 Organochlorines

Saka (2004) analysed teratogenic effects through FETAX for three major DDT metabolites. For the *para-para* isomers of DDT, DDD and DDE assayed, the results showed sensitivity of *X. laevis* both in terms of mortality and malformations in the order of *p,p*-DDD > *p,p*-DDT > *p,p*-DDE, which happens to be the inverse of their expected half-lives in the environment. The LC50 of *p,p*-DDT was 35.8 mg/L with the EC50 at 14.71 mg/L. The LOEC for malformations was 6.24 mg/L, but the minimum concentration to inhibit growth (MCIG) was 1.5 mg/L. This places the MCIG/LC50 ratio at 0.042. The teratogenic index (TI = LC50/EC50) value for *p,p*-DDT was 2.4. As a standard in FETAX, any compound with a MCIG/LC50 ratio lower than 0.3 and a TI above 1.5 is considered teratogenic.

3.1.2 Pyrethroids

Teratogenicity of deltamethrin was analysed by Channing (1998), who reported an LC50 and EC50 of 0.19 mg/L and 0.006 mg/L, respectively, with a resulting TI of 31. Even though Channing (1998) did not report the MCIG or LOEC, a TI value 20 times higher than the 1.5 cut-off means that deltamethrin is certainly regarded as teratogenic.

3.1.3 Organophosphates

Malathion was the only compound for which FETAX results were reported from multiple sources. Snawder and Chambers (1989) showed malathion to have an LC50 of 10.9 mg/L. They reported the EC50 in terms of specific malformations, the lowest of which was an abnormal pigmentation EC50 of 0.33 mg/L and the highest an abnormal notochord EC50 of 2.16 mg/L. Bonfanti et al. (2004) did not expose tadpoles to a high enough range of malathion to calculate an LC50, with their highest concentration of 6 mg/L showing mortality of only 12%. They did, however, report an EC50 of 2.39 mg/L for total malformations. This value is similar to the notochord abnormality EC50 of Snawder and Chambers (1989), as Bonfanti et al. (2004) reported the malformations observed to consist mostly of abnormal tail flexure, which is relatable to abnormalities of the notochord.

Neither Channing (1998) nor Snawder and Chambers (1989) reported a lowest-observed-effects concentration (LOEC) or minimum concentration to inhibit growth (MCIG) to compare with results from Saka (2004).

Technical DDT consists of about 85% *p,p*-DDT and about 15% *o,p*-DDT (ATSDR 2002) with variable traces of other forms such as *p,p*- and *o,p*-DDD and sometimes *o,o*-DDT. The majority of risk from direct exposure is therefore related to *p,p*-DDT. Hence, in analysing initial toxicity from MVC application, acute toxicity and teratogenic risk to *Xenopus* embryos, when exposed to the equal amounts of each MCVP, would be greatest from deltamethrin n, followed by malathion and lastly DDT. However, these compounds are not necessarily used in equal amounts during MVC. With DDT being banned under the Stockholm convention (UNEP 2009) from other forms of use in most countries, it also carries the lowest risk of being used for other purposes in the environment. Malathion and deltamethrin are, however, actively used in agriculture, in turn increasing risk to non-target organisms, including *X. laevis*, in aquatic environments. Deltamethrin further has the highest potential to cause developmental malformations at concentrations that would not cause significant mortality, based on its high TI value.

Several other studies included in this review followed the FETAX protocol in terms of exposure but reported on different endpoints to that of the standardised FETAX format, as in the previously described studies. These studies were therefore included in the acute toxicity or sub-lethal and chronic toxicity data collections (Tables 4 or 5), based on exposure duration and endpoints measured.

3.2 Acute Toxicity

Given greater variability in methods, age of animals and report structure outside the FETAX parameters, there were a few more acute toxicity studies that met the requirements for inclusion in this review. Acute toxicity records could be found only for deltamethrin, cyhalothrin (CAS: 91465-08-6), cypermethrin, fenitrothion (CAS: 122-14-5) and malathion (Table 4).

3.2.1 Pyrethroids

Aydin-Sinan et al. (2012) used a modified FETAX protocol to expose *X. laevis* Nieuwkoop-Faber (NF) stage 46 tadpoles to deltamethrin in a commercial formulation. The concentration range of deltamethrin (1.1–38.4 μg AI/L) was too low to induce 50% mortality at 96 h. Maximum concentration induced 21.7% mortality, which is within reasonable expectations compared to the 190 μg/L LC50 (also 96 h) reported by Channing (1998) on embryos using the FETAX method. A 168 h LC50 reported by Aydin-Sinan et al. (2012) is 6.26 μg AI/L.

Aydin-Sinan et al. (2012) also exposed *X. laevis* to λ-cyhalothrin in commercial formulation which showed greater toxicity than deltamethrin. The 96 h exposure also did not induce 50% mortality, but at 10 μg AI/L as the maximum concentration, 46.7% of the test population died. The 168 h LC50 was reported as 3.94 μg AI/L.

Yu et al. (2013) analysed the toxicity of α-cypermethrin to *X. laevis*. Their study showed *X. laevis* to be one of the most sensitive anurans to α-cypermethrin based on literature comparisons, and it was suggested as a useful species for studying the effects of pyrethroids. Very little literature on other anurans was available for comparison at the time of publication, which should be taken into account. The LC50 for *X0. laevis* to α-cypermethrin was recorded as 30.6 (CI 27.1–34.7) μg/L and

Table 4 Summary of results from studies that performed non-FETAX acute toxicity tests on *Xenopus laevis* with regard to any of the 12 WHO-recommended MVCPs

Pesticide	Life stage	Exposure range (mg/L)	Temperature (°C)	LC50 24 h	LC50 96 h	LC50 168 h	Notes	Source
Fenitrothion	Embryo	0.1–10	18 25 30	18– 25 °C > 10 30 °C = 0.33	NA	NA	EC50 at 18 °C = 4.2; 25 °C = 3.7; 30 °C = 0.17 (abnormalities leading to post-hatching death)	Elliot-Feeley and Armstrong (1982)
Deltamethrin	Larvae (NF 46)	1.1×10^{-3}–38.4×10^{-3}	23	NA	>38.4×10^{-3} (21.7% mortality)	6.26×10^{-3}	AI-based concentration in commercial formulation	Aydin-Sinan et al. (2012)
λ-Cyhalothrin	Larvae (NF 46)	1.1×10^{-3}–10×10^{-3}	23	NA	>10×10^{-3} (46.7% mortality)	3.94×10^{-3}	AI-based concentration in commercial formulation	Aydin-Sinan et al. (2012)
Malathion	Embryo	0.5–8.0	22–24	NA	5.4 (4.9–6.0)	NA		Yu et al. (2013)
α-Cypermethrin	Embryo	10×10^{-3}–0.16	22–24	NA	30.6×10^{-3} (27.7×10^{-3}–34.8×10^{-3})	NA		Yu et al. (2013)
Malathion	Larvae (NF 46)	1.3–10	22–24	NA	6.76 (6.11–7.52)	NA		Yu et al. (2013)
α-Cypermethrin	Larvae (NF 46)	1.25×10^{-3}–20×10^{-3}	22–24	NA	6.9×10^{-3} (5.7×10^{-3}–8.5×10^{-3})	NA		Yu et al. (2013)
α-Cypermethrin	Larvae (NF47/48)	0.5–5	22–23	NA	NA	NA	Combined UVB exposure increased mortality but was similar to UVB alone	Yu et al. (2015b)
Malathion	Larvae (NF47/48)	0.5×10^{-3}–5×10^{-3}	22–23	NA	NA	NA	Combined UVB exposure had no mortality interaction	Yu et al. (2015b)

Route of exposure for all studies in this table was immersion. *NA* not applicable

6.9 (CI 5.7–8.5) µg/L for embryos and free-swimming larvae, respectively, making α-cypermethrin more toxic to free-swimming larvae. This result brings into question the use of a single method such as FETAX as a definitive sensitivity test for these compounds to anurans. Incorporation of other life stages seems necessary for a holistic approach to assess the effect these chemicals have on anuran populations and community structure. Publications by Yu et al. (2013, 2015b) were the only records of α-cypermethrin acute toxicity testing on *Xenopus* available for the present review. Yu et al. (2015b) investigated combination effects of simultaneous exposure to α-cypermethrin and ultraviolet-B radiation (UVB). The addition of UVB (intensity of 22–28 µW/cm^2) showed increased mortality, but this was mostly attributed to the toxic effects of UVB itself, and no discernible relationship between pesticide-caused mortality and UVB-caused mortality could be distinguished. The results did however indicate a synergistic effect between the two in terms of sub-lethal effects such as growth and malformations.

3.2.2 Organophosphates

The only study on acute toxicity listed in Pauli et al. (2000) involving *Xenopus* and an MVCP was Elliot-Feeley and Armstrong (1982) involving *X. laevis* and fenitrothion. This was also the only existing acute toxicity record of fenitrothion on *Xenopus* we could find. The authors exposed *X. laevis* embryos at gastrulation phase to a range of concentrations (0.1–10 mg/L) of fenitrothion for 24 h at various temperatures. The results showed an LC50 above 10 mg/L at 18 °C (8% mortality at 10 mg/L) and 25 °C and an LC50 of 0.33 mg/L at 30 °C. At 18 °C, when exposed to a saturated solution (estimated at 30 mg/L), mortality increased to 89%, which places the theoretical LC50 at 18 °C somewhere between 10 and 30 mg/L. The surviving embryos were raised to NF stage 37 and morphological changes recorded. The authors classed tadpoles based on the severity of morphological change, which showed increased severity corresponding to increased concentration. Significant differences in morphology were observed at 1 mg/L (at 18 °C), while increased mortality from the control was measured at only 10 mg/L. Sensitivity tests were also conducted with 10 mg/L exposures at 25 °C which indicated the highest mortality when exposed during stage 11–15 but with the highest incidence of morphological changes when exposed at stage 8–11. The sensitivity differences in terms of mortality at different developmental stages once again (as with α-cypermethrin) seem to indicate that a single life-stage exposure is inadequate to determine realistic toxicity information when working with anurans.

Yu et al. (2013) also investigated the toxic effects of several pesticides, including malathion, following the FETAX method, except for a few modifications such as not de-jellying the eggs. This provided a more realistic exposure scenario but excluded this study from outright comparison to other FETAX studies. Other parameters they adjusted were the mixing of egg clutches, larger test volumes, and tadpoles were considered dead when they did not respond to prodding rather than when they had no heartbeat (as described in the latest version of FETAX; ATSDR 2012). The results from Yu et al. (2013) showed malathion as having an LC50 of 5.4 (95% CI 4.9–6.0) mg/L and 6.7 (95% CI 6.1–7.5) mg/L for stage-8–11 and stage-46 larvae, respectively. This is lower than the values measured by Snawder and Chambers (1989) and Bonfanti et al. (2004), which the authors ascribed to their alterations on the FETAX protocol. In Yu et al. (2013), malathion showed less acute toxicity (endpoint: mortality) to free-swimming larvae (NF stage 46) than to embryos (NF stages 8–11) with similar sensitivity indicated by growth and malformation endpoints. The authors also compared the sensitivity of *X. laevis* larvae to values from the literature on other amphibian species with *X. laevis* showing median sensitivity to malathion. Yu et al. (2015b) also investigated additive effects of UVB radiation to malathion toxicity but with no discernible interaction in terms of mortality. However at lower concentrations, the addition of UVB significantly increased the incidence of malformations such as abnormal tail tip and gut observed in conjunction with malathion exposure. Oedema incidence, caused by UVB alone, was reduced in combined exposure with higher concentrations of malathion.

3.2.3 Relative Toxicity of MVCPs Towards *Xenopus*

In order to evaluate the sensitivity of *Xenopus* sp. (data only available for *X. laevis* as representative of *Xenopus* sp.) to MVCPs, we analysed the acute toxicity data above to rank the 12 WHO MVCPs based on their relative potency towards *Xenopus* sp. We utilise the term potency in a generic manner to indicate comparative toxicity towards *Xenopus* sp. rather than terms such as hazard or risk since in the context of risk assessment these terms would imply so form of known relationship between exposure and effect. By plotting the available LC50 values for each pesticide vs the recommended manufacturer's application rate (m^2), we derived course relative MVCP potency categories (Fig. 1). High potency towards *Xenopus* sp. relates to a high toxicity (i.e. low LC50 value) and a high application rate, and conversely the lowest application rate and highest LC50 value would result in the lowest potency. The boundary between high and low potency was derived from the combination of highest application rate between all compounds and highest LC50 between all compounds as the best case for toxicity and worst case for application. This combination then provides a separation point for compounds having worse or better combination outcomes than the combination of worst and best case scenarios. Any points below this line are considered moderate to high risk, while any points above are considered low to very low risk. Zero risk cannot be used as a predicted value if application is a non-zero value. Based on these combinations, relative potency values were calculated using the following formula: Log((average dose $\times$ 1/mean LC50) + 1) (Fig. 2).

The use of cyhalothrin for MVC theoretically contributes the greatest acute toxicity potency to *Xenopus* sp. in a natural setting during MVC application out of all the MVCPs for which toxicity data is available. Fenitrothion, deltamethrin and malathion also have high relative potency, while DDT has a moderate relative potency. α-Cypermethrin poses the lowest risk for acute toxicity to *Xenopus* sp. out of the possible MVCPs. The lack of acute toxicity data for bendiocarb (CAS: 22781-23-3), bifenthrin (CAS: 82657-04-3), cyfluthrin (CAS: 68359-37-5), etofenprox (CAS: 80844-07-1), pirimiphos-methyl (CAS: 29232-93-7), and propoxur (CAS: 114-26-1) means that 50% of the WHO MVCPs could not be evaluated in this manner.

It should be stressed that there are specific limitations to the use of the relative potency values calculated in this manner. Firstly factors such as frequency of use, distance from spraying, temperature, wind, soil composition, runoff from precipitation, photolysis and other forms of degradation will all affect the extent to which these animals are essentially exposed in the wild. These factors are not included as most of them are going to be scenario specific and using theoretical values can provide even more misleading results. The calculations performed in this review merely compare the potential to be introduced in a broad sense vs the potential to have a toxic effect on these animals. This provides comparative potency between the pesticides discussed in this review only. The lack of toxicity data on many of the discussed pesticides also limits the interpretation of these results in the wider context of all 12 WHO-recommended MVCPs. In fact these limitations on the meta-analysis already provide insight into the issues regarding lack of data in this field. That being said, the comparisons provided between these pesticides from the relative potency values provide rationale as to prioritising pesticides for future research.

Pyrethroids are the compound group that contribute the greatest relative potency to *Xenopus* sp. in terms of having the potential to cause frog mortality in quantities used for MVC in Africa (Fig. 2). This is due to the high sensitivity *Xenopus* has towards these compounds compared to other MVCPs. Even with DDT (organochlorine) application rate 26 times higher than the mean application rate of the pyrethroids, the mean relative potency of the three pyrethroids analysed is still higher than that of DDT. Based on application rates and sensitivity, we can discern that λ-cyhalothrin holds the greatest potential for acute toxicity to *Xenopus* sp. followed by deltamethrin. However, all the toxicity tests included were not conducted under the exact same conditions, lowering the confidence of such a conclusion.

The use of multiple life stages in anuran acute toxicity testing plays an important role in amphibian ecotoxicology as the most sensitive life stage was not consistent throughout reviewed studies where both embryo and larval stages were used. The most sensitive life stage appears to be compound-specific and may be related to the specific mechanism of action of each pesticide. Another issue already touched on with the available toxicity data is the lack of structure. Currently three internationally standardised tests for the *Xenopus* model are available: the Organisation for Economic Co-operation and Development (OECD) chemical safety testing protocol No. 231, amphibian metamorphosis assay (AMA); No. 241, larval amphibian growth and development assay (LAGDA); and FETAX. The AMA is to some extent a 21-day variation of FETAX and focuses only on the thyroid axis effects compounds can have that affect the metamorphoses of frogs. It can be applied to different anurans but was also specifically designed around *X. laevis*, like FETAX. The LAGDA stemmed from the AMA and is considered a more comprehensive toxicity analysis enabling measurement of endocrine disruption through various pathways.

The issue with using either of these assays for other species comes down to the developmental staging of the frogs. General frog development stages, as defined by Gosner (1960), have less detail than NF staging (Nieuwkoop and Faber 1994) for *X. laevis*, which is what makes the generalisation possible, but limits quantification during assays and comparability to NF staged studies. The aquatic-terrestrial range of amphibians also provides complications as not all amphibians can be immersed continuously in exposure media; some species will die if they have prolonged contact with water due to unique osmoregulation adaptations. This is the same uniqueness that makes frogs so habitat-specific. Studying frog ecotoxicology in a laboratory is one thing, but relating those results back to real-world exposure scenarios raises complications in itself between species and the respective habitats they inhabit.

The lack of acute toxicity data for WHO-recommended pesticides on a standard model organism is alarming. As *Xenopus* is not the only standard testing frog species in the world, it is fitting also to compare literature on other amphibian species. Species sensitivity distributions (SSDs) were compiled from the LC50 values discussed in this review and US-EPA ECOTOX database results on amphibians in relation to the 12 WHO MVCPs in question. Figure 3 shows an SSD of amphibians to malathion. LC50 data for studies ranging between 48 and 96 h was included in the analysis. Data fitting these criteria were found for nine amphibian species. *Xenopus laevis* is shown to be the second least susceptible species in this regard. SSDs were also calculated for deltamethrin, DDT and cypermethrin in the same manner (data not shown); however, there were less than seven data points available for each of these pesticides, increasing the error margin severely. Both DDT and deltamethrin

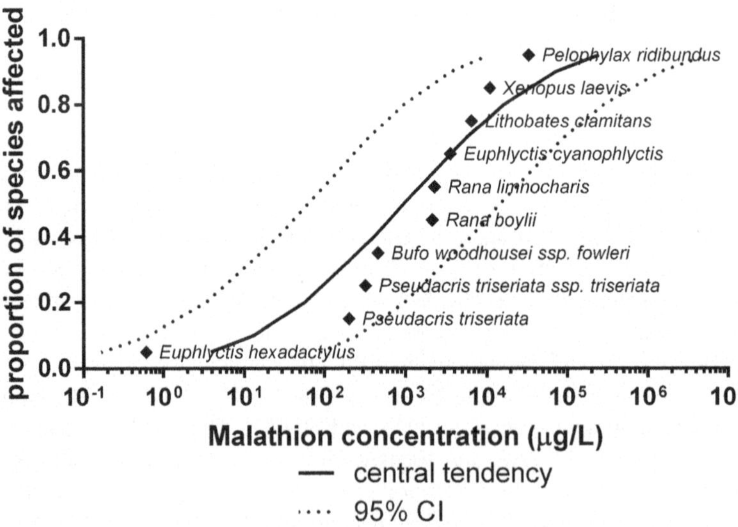

Fig. 3 Anuran species sensitivity distribution for malathion showing the proportion of species affected vs LC50 concentration. Both 48 h and 96 h LC50 data were included

showed *X. laevis* to be the least susceptible species, but for cypermethrin *X. laevis* was the second most susceptible species, with only the Asian common toad *Duttaphrynus melanostictus* having a lower LC50. Interestingly, in terms of cypermethrin, *X. laevis* is thus a relatively sensitive anuran, but this pesticide was shown to have the lowest relative potency value to *X. laevis*. Conversely, malathion and deltamethrin had fairly high potency values compared to other MVCPs but showed *X. laevis* as tolerant, compared to other frog species. This indicates an inflated potency towards most other frog species for these two compounds. Exposure scenarios for other species would differ based on their preferred habitats, which may lower the realistic risk posed by these compounds. The lack of data on other anuran species is also alarming as it indicates *Xenopus* species aren't the only frogs neglected in this topic of research. Larger and more relevant exposure and sensitivity datasets are required in order to be able to properly investigate risks these pesticides pose to frogs. Realistic (or simulated-realism) exposure-response and site-specific application-risk investigations into all MVCPs, not only for *X. laevis* but also other frog species, seems to be a logical and crucial next step for amphibian ecotoxicology research and conservation in MVC regions. Of all the MVCPs with data available, the compound with the highest relative potency towards *X. laevis* was λ-cyhalothrin. The only other publication on acute toxicity to frogs for this compound in the ECOTOX database is for the Bog Frog *Fejervarya limnocharis*, previously *Rana limnocharis*, which showed a 48 h LC50 of 4 μg/L (Pan and Liang 1993), compared to the 3.97 μg/L LC50 for *X. laevis* only after 3.5 times longer exposure. The lack of comparable toxicity data with regard to MVCPs and frogs in general raises the question of how we can be sure that the current use of pesticides for MVC, excluding agricultural use, is not causing permanent damage to frog populations in malaria-risk regions.

It is unlikely that mass mortalities of frog populations due to MVC may have gone unnoticed, but even if exposure does not result in mortality, it is conceivable that sub-lethal effects may occur. In the next section, we summarise literature available on sub-lethal and chronic effects of MVCPs on *Xenopus* sp.

3.3 Sub-Lethal and Chronic Effects

3.3.1 Organochlorines

DDT is a well-known endocrine-disrupting compound (EDC) (Hoffmann and Kloas 2016). Palmer and Palmer (1995) investigated the use of plasma vitellogenin as a biomarker for oestrogenic activity in *X. laevis* males exposed to DDT (Table 5). When exposed to 1 μg/g and 250 μg/g o,p-DDT through injection for 7 days, with blood analysed on day 14, both concentrations showed detectable vitellogenin levels significantly higher than control and increasing in a dose-dependent manner. Hoffmann and Kloas (2016) studied the EDC effects of DDE on male *X. laevis* through analysing call behaviour. Significant reduction in

Table 5 Studies performed on *Xenopus laevis* with regard to any of the 12 WHO-recommended MVCPs investigating sub-lethal or chronic exposure effects

Pesticide	Life stage	Exposure route	Temperature (°C)	Exposure time (days)	Exposure concentration	Notes	Source
o,p-DDT	Adult (males)	Injection	21	7	1 and 250 (μg/g)	Injected for 7 days but analysed at day 14. Vitellogenin was extractable from the plasma of treated frogs, but it was not extractable from the plasma of control specimens	Palmer and Palmer (1995)
p,p-DDE	Adult (males)	Immersion	21.6	4	3.18×10^{-6} & 3.18×10^{-4} (mg/L)	3.18 ng/L showed estrogenic activity, while 318 ng/L showed antiandrogenic activity	Hoffmann and Kloas (2016)
λ-Cyhalothrin	Tadpoles (NF 46)	Immersion	23	7	5.5×10^{-4}–0.01 (mg AI/L)	Decreased acid phosphatase, aspartate aminotransferase, glutathione-*S*-transferase and lactose dehydrogenase	Aydin-Sinan et al. (2012)
Deltamethrin	Tadpoles (NF 46)	Immersion	23	7	1.1×10^{-3}–0.0384 (mg AI/L)	Decreased acid phosphatase and aspartate aminotransferase	Aydin-Sinan (2012)
α-Cypermethrin	Tadpoles (NF 49)	Immersion	20	14	0.001–0.025 (mg/L)	0.015–0.025 mg/L showed clastogenic activity and increase in number of micronucleated red blood cells	Rudek and Rożek (1992)
α-Cypermethrin	Tadpoles (NF 47)	Immersion	22	9	2.4×10^{-4} (mg/L)	Increased heat shock protein 70 and interleukin-1β gene expression	Martini et al. (2010)
α-Cypermethrin	Tadpoles (NF 46)	Immersion	22–23	4	5.0×10^{-4} (mg/L)	No light (UVB) avoidance behaviour influences were observed	Yu et al. (2014)
α-Cypermethrin	Tadpoles (NF 47)	Immersion	22–23	4	0.005 (mg/L)	Increased DNA damage. Combined with 63 mW/m^2 UVB for last 7 h showed decrease in most of this damage	Yu et al. (2015a)
Malathion	Embryos	Immersion	20	4	0.1–100 (mg/L)	Dose-dependent reduction in NAD^+	Snawder and Chambers (1989)

Malathion	Embryos	Immersion	23	4	1, 5 and 10 (mg/L)	Lysyl oxidase inhibition (IC50: 2.3×10^{-4} mg/L) and proline hydroxylase inhibition (IC50: 0.2 mg/L) Less collagen fibres observed, but total collagen quantification showed no reduction	Snawder and Chambers (1993)
Malathion	Embryos	Immersion	23	4	1, 5 and10 (mg/L)	Dose-dependent decrease in length. Dose-dependent increase in abnormal pigmentation, abnormal gut and notochordal bending. Dose-dependent reduction of intestinal loops with a concurrent increase in anterior intestine diameter	Snawder and Chambers (1990)
Malathion	Tadpoles	Immersion		30	1.0×10^{-9}–1.0 (mg/L)	More than 50% mortality at 30 days in 1.0 mg/L. AChE reduced	Webb and Crain (2006)
Malathion	Embryos	Immersion		3	1.0 and 2.5 (mg/L)	Time-dependent increase in notochord axis angle and significant reduction in length at both concentrations	Chemotti et al. (2006)
Malathion	Tadpoles (NF 46)	immersion	22–23	4	1.0×10^{-3} & 5.0×10^{-3} (mg/L)	No light (UVB) avoidance behaviour influences were observed	Yu et al. (2014)

advertising calls (female-attracting call) were seen at 3.18 ng/L exposure. This reduction, however, was not significant at 318.0 ng/L exposure, which did in turn show a significant increase in the amount of ticking (a type of release call). The manner in which calls were altered by the different concentrations of DDE is, according to the authors, suggestive that both estrogenic and antiandrogenic mode of action can be elicited at different concentrations with lower concentrations showing estrogenic activity and higher concentrations showing antiandrogenic activity.

3.3.2 Pyrethroids

Apart from acute toxicity, Aydin-Sinan et al. (2012) also analysed a set of biomarker responses to analyse the sub-lethal effects of deltamethrin and λ-cyhalothrin on *X. laevis*. Both compounds significantly decreased ACP (acid phosphatase) and AST (aspartate aminotransferases) after 24 h exposure. Cyhalothrin also significantly decreased CaE (carboxylesterase) and GST (glutathione-*S*-transferase), with LDH (lactate dehydrogenase) being inhibited by the highest concentration (10 μg AI/L). To our knowledge, this publication is the first report on the use of biomarker responses with *Xenopus* sp. in relation to MVCPs.

Rudek and Rozek (1992) exposed *X. laevis* to cypermethrin between 1 and 25 μg/L and found a dose-dependent increase in the occurrence of erythrocytes with micronuclei. The increase was, however, not as severe as in *Rana temporaria*. Martini et al. (2010) exposed *X. laevis* tadpoles to 0.24 μg/L α-cypermethrin in order to measure immune responses. Cypermethrin was found to increase the messenger-ribonucleic acid (m-RNA) expression of heat shock protein 70 and interleukin-1β. Both of these play important roles in immune response, and alterations to their normal expression could result in a compromised response to influences such as disease. In Yu et al. (2014) the behaviour of tadpoles exposed to pesticides was measured in terms of UVB radiation avoidance. A 0.5 μg/L cypermethrin exposure, however, showed no difference in time spent in or outside of the UVB-exposed area compared to control. A 1 μg/L cypermethrin exposure was excluded from the results as axial malformations interfered with swimming behaviour and subsequently light avoidance behaviour. Yu et al. (2015a) exposed *X. laevis* embryos to cypermethrin and UVB simultaneously looking at sub-lethal effects in the form of deoxyribonucleic acid (DNA) damage. Cyclobutane pyrimidine dimer accumulation, indicating DNA adducts, increased significantly compared to both control and UVB-only exposure, at 5 μg/L cypermethrin exposure for 96 h with 63.0 mW/m^2 UVB introduced for the last 7 hours of exposure. The m-RNA expression of genes indicative of DNA damage showed cypermethrin alone down-regulated the xeroderma pigmentosum group A (XPA) gene and upregulated the xeroderma pigmentosum group G (XPG) gene and Cockayne syndrome A (CSA) gene that play a role in transcription-coupled DNA excision repair, and cypermethrin also upregulated expression of the Muir-Torre syndrome (MUTL) gene which is involved in mismatch repair. In combination with UV, however, most of these effects were nullified, with only XPG up-regulation being enhanced even further. These results indicate that sub-lethal concentrations of cypermethrin can have stress response effects in *X. laevis* and that some of these effects may be enhanced in a natural environment with combined exposure to UVB radiation.

3.3.3 Organophosphates

In addition to mortality and developmental malformation, Snawder and Chambers (1989) also investigated the effects of malathion on nicotinamide adenine dinucleotide (NAD^+) in the exposed *X. laevis* embryos. They found significant reduction of NAD^+ showing a gradual almost linear decline on a log (dose) vs response curve. This dose-related reduction in NAD^+ is deemed independent from malformations observed, as other compounds tested showed variation in malformations observed with similar reduction in NAD^+. Following these results, Snawder and Chambers published two more studies (Snawder and Chambers 1990, 1993) in which they also exposed *X. laevis* embryos to malathion using the FETAX protocol. In Snawder and Chambers (1990), the effect of tryptophan combined with malathion on the NAD^+ levels was tested, and, based on malformations observed, it was concluded that tryptophan could effectively combat the NAD^+ reduction effects of malathion but with no resulting change in the malformations observed from malathion exposure. The authors also showed that 4-day-old embryos exposed for 1 day showed similar malformations to those exposed for 4 days from 0 days old, making the 4th day of development the most critical point in development in terms of malformation. Snawder and Chambers (1993) focused on the effects of malathion on collagen and subsequent notochord bending. Ultrastructural examination showed fewer collagen fibres in the notochords of exposed embryos, but quantification of collagen in homogenised embryos did not differ significantly from controls. Malathion was, however, shown to inhibit lysyl oxidase and collagen proline hydroxylase activity in *X. laevis* embryos in a dose-dependent manner, with IC50s of 0.7 nM and 0.58 μM, respectively. The inhibition prevents triple helix collagen from forming and in turn reduces the extracellular collagen fibres, even though total collagen content assays show no reduction.

Webb and Crain (2006) exposed *X. laevis* tadpoles to different concentrations of malathion over 30 days, with 1.0 mg/L showing a significant mortality rate compared to the control and with more than 50% mortality at the end of the experiment. Webb and Crain (2006) exposed 3-week-old tadpoles, but the developmental stages were not reported, only that stages did not differ between treatments. This makes comparison difficult. Chemotti et al. (2006) investigated the teratogenic effects of malathion on *X. laevis* embryos using a modified 72 h FETAX method deigned for practical teaching applications. The tadpoles showed significant time-dependent increase in the notochord axis angle over 72 h at 1.0 and 2.5 mg/L malathion exposure, together with a significant reduction in length at 72 h exposure for both concentrations, compared to controls. Yu et al. (2014) also tested UVB avoidance behaviour in relation to malathion exposure in *X. laevis* tadpoles, but neither 1 nor 5 µg/L exposures showed significant differences in behaviour.

3.4 In Vitro Studies

Due to *X. laevis*' extensive use in teaching and as a physiology model, many physiological tests have been performed on tissue preparations and cell cultures of these animals and of their neuromuscular system in particular, in order to better understand the mechanistic action of neurotoxic compounds. Such studies provide useful information with regard to the mechanisms of action of pesticides. These studies are listed in Table 6.

3.4.1 Organochlorines

Van den Bercken et al. (1973b) investigated the action potential of muscle nerve fibres of *X. laevis* exposed to 25 mg/kg *p,p*-DDT. The poisoning symptoms were described as restlessness and hyper-excitability, which increased in intensity until convulsions occurred, coupled with excessive skin mucus secretion. After euthanasia the action potential was measured on excised muscles. The treated animals showed an increase in the amplitude of negative after-potential. The duration of action potential was also increased in treated animals, attributed to a slower falling phase. Van Den Bercken et al. (1973a) investigated the effects of DDT on prepared *X. laevis* lateral line organs at different temperatures. DDT exposure (2–4 mg/L) caused trains of impulse spikes instead of the normal single-impulse spontaneous activity. The number of spikes per train increased at lower temperatures, while the frequency of trains increased at high temperatures. Åhrem and Frankenhaeuser (1974) measured the permeability effects of DDT and its metabolites on *X. laevis* nerve fibres using potential clamping. Effects seen with *Lithobates pipiens* (previously *Rana pipiens*) from DDT on the permeability of Na^+ could not be replicated with *X. laevis*. Dichlorodiphenylacetic acid (DDA; CAS: 83-05-6), a minor metabolite of DDT, was found to increase the permeability of K^+ through the nerve fibres at 4 mg/L. In a study by Århem et al. (1974), the acute effects of the same DDT metabolites were investigated on myelinated nerve fibres of *X. laevis*. This publication seems to be, to some extent, a detailed summary of significant results from Århem and Frankenhaeuser (1974) and merely confirms that the differences between *X. laevis* and *Lithobates pipiens*, in terms of DDT exposure mentioned in Århem and Frankenhaeuser (1974), are in fact due to differences in species response, along with a more detailed description of how DDA affects the K^+ membrane potential through interfering with inactivation of the potassium-gated channel. Repetitive firing in the lateral line organ of *X. laevis* caused by DDT exposure was also demonstrated by Akkermans et al. (1975). Vijverberg et al. (1982b) found the mechanism of action of DDT to be related to opening of the sodium channels in *Xenopus* myelinated nerve

Table 6 In vitro studies performed on *Xenopus laevis* with regard to any of the 12 WHO-recommended MVCPs

Pesticide	Tissue exposed	Temperature (°C)	Exposure concentration	Notes	Source
p,p-DDT	Muscle nerve fibres	21–25	25 (mg/kg)	Adult frogs dosed in vivo and nerve fibres analysed after death. Increase in amplitude of negative after-potential and slowing down of the falling phase of action potential without affecting resting potential; no repetitive activity	Van den Bercken et al. (1973b)
DDT	Lateral line organ	4–35	2–4 (mg/L)	Caused impulse spike trains number of spikes per train correlated negatively with temperature and frequency of trains correlated positively with temperature	Van den Bercken et al. (1973a)
DDA	Nerve fibres	12	4 (mg/L)	DDT was shown to affect Na^+ permeability in nerve fibres of *Rana pipiens*, but similar results could not be duplicated with *Xenopus laevis* nerve fibres. DDA caused increased permeability of of K^+ at 4 mg/L	Århem and Frankenhaeuser (1974)
DDA	Nerve fibres	12	4 (mg/L)	DDA was shown to affect the K^+ membrane potential through interfering with inactivation of the potassium gated channel	Århem et al. (1974)
DDT	Lateral line organ	19–21	2–5 (mg/L)	Repetitive activity recorded after 2–5 mg/L exposure	Akkermans et al. (1975)
DDT	Myelinated nerve fibres	15	3.5–14 (mg/L)	Modified sodium channel gating selectivity reduced the rate of closing of the activation gate in myelinated nerve fibre. Similar results to non-cyano pyrethroids	Vijverberg et al. (1982a)
p,p-DDT	Primary liver cells	4	3.5×10^{-4}–354.5 (mg/L)	Low oestrogen binding affinity, only compound tested with lower binding affinity than endogenous steroids tested	Lutz and Kloas (1999)
α-Cypermethrin	Lateral line organ	8–22	2.1 (mg/L)	Repetitive activity was only induced in in vivo exposed lateral line organs. Impulse train lengths reduced as temperature increased	Vijverberg and Van den Bercken (1982)
α-Cypermethrin	Myelinated nerve fibres	0–25	0.42–8.3 (mg/L)	Significantly reduced the rate of relaxation of sodium channel activation gate	Vijverberg et al. (1983)

(continued)

Table 6 (continued)

Pesticide	Tissue exposed	Temperature (°C)	Exposure concentration	Notes	Source
α-Cypermethrin	Muscle-nerve preparation	18	0.42–41.6 (mg/L)	No repetitive activity with respect to end-plate potentials was observed; however, up to 60 action potentials occurred	Ruigt and Van den Bercken (1986)
Deltamethrin	Myelinated nerve fibres	19–21	0.5 (mg/L)	Single fibre experiments were performed at 15 °C, Deltamethrin induced frequency dependent reduction in action potential at 10^{-6} M	Vijverberg and den Bercken (1979)
Deltamethrin	Lateral line organ	8–22	0.5–2.5 (mg/L)	Deltamethrin induced repetitive activity in lateral line organs at all concentrations through in vitro exposure. Impulse train lengths reduced as temperature increased	Vijverberg and Van den Bercken (1982)
Deltamethrin	Myelinated nerve fibres	0–25	0.5–10.1 (mg/L)	Significantly reduced the rate of relaxation of sodium channel activation gate. Showed slowest decay of tail currents out of 11 pyrethroids analysed	Vijverberg et al. (1983)
Deltamethrin	Muscle-nerve preparation	18	0.5–50.5 (mg/L)	No repetitive end-plate potentials for R-cis, S-cis and trans, and R-trans and R-cis were able to produce up to 60 action potentials	Ruigt and Van den Bercken (1986)
Des-cyano-deltamethrin	Muscle-nerve preparation	18	15.2 (mg/L)	15–30 end-plate potentials and 2–18 action potentials	Ruigt and Van den Bercken (1986)
Malathion	Blood plasma	4	2.6 (mg/L)	Zero to weak binding interference shown by malathion between thyroid hormone receptor binding action of endogenous hormones	Yamauchi et al. (2002)
Malathion	Oocytes	NR	0.1 (mg/L)	Reduced oocyte maturation time significantly. Effect was pronounced in 1:1 malathion:atrazine mixture exposures at both 0.05 and 0.1 mg/L	Ji et al. (2016)

fibres, with DDT having no major effect on depolarisation of sodium channel but causing sodium tailing after repolarization. Lutz and Kloas (1999) measured the oestrogen binding activity in primary liver cell cytosol from *X. laevis*. Between seven endogenous steroids and seven exogenous ligands known for oestrogen receptor binding analysed, *p,p*-DDT had the lowest affinity for oestrogen receptor binding and was the only exogenous ligand with lower binding affinity than the endogenous steroids measured.

3.4.2 Pyrethroids

Vijverberg and Van den Bercken (1982) studied the effect of pyrethroids on the lateral line organs of *X. laevis*. Cypermethrin failed to induce repetitive activity within the first 5 h in in vitro-exposed studies. In in vivo-exposed lateral line organs, however, long-lasting trains of nerve impulses were induced after 4 h when exposed to 5×10^{-6} M cypermethrin. In a study by Vijverberg et al. (1983) on myelinated nerve fibres, Na^{+} tail currents were caused by both cypermethrin and deltamethrin with deltamethrin showing the slowest decay of tail currents out of 11 pyrethroids tested. The rate of decay accounted for the differences in repetitive activity induced by various pyrethroids. Ruigt and Van den Bercken (1986) tested the effects of pyrethroids on end-plate potentials and muscle action potentials in excised pectoralis nerve-muscle preparations of *X. laevis*. At 100 μM cypermethrin failed to induce repetitive end-plate potentials. Cypermethrin was, however, very effective at producing repetitive action in muscle fibre.

In a study on the effects of deltamethrin (referred to as decamethrin in that study), Vijverberg and Van den Bercken (1979) noted that cypermethrin causes frequency-dependent reduction of action potential in *X. laevis*. Deltamethrin was shown to induce this same depression at 10^{-6} M. Vijverberg and Van den Bercken (1982) also analysed the effect of deltamethrin on the lateral line organ. Long-lasting trains of nerve impulses were observed in in vitro experiments. The number of impulses per train was significantly higher in α-cyano-containing pyrethroids (deltamethrin, cypermethrin, etc.) than in non-cyano-containing pyrethroids (permethrin, etc.). All pyrethroids analysed showed a temperature-dependent reduction in the length of impulse trains induced. Deltamethrin exposure also did not induce repetitive activity in excised sciatic nerves at 10^{-5} M concentrations after 24 h. Ruigt and Van den Bercken (1986) also found no repetitive end-plate potentials for deltamethrin exposure. End-plate potentials were, however, induced by des-cyano-deltamethrin (deltamethrin without its α-cyano group). Muscle action potentials were induced by both forms.

In general, mechanisms of action derived from the above studies showed activity between DDT, and pyrethroids were shown to be very similar, with similar sodium channel activity suggested. Slight differences between different compounds can mostly be attributed to relaxation times after action (Vijverberg et al. 1982b).

3.4.3 Organophosphates

Yamauchi et al. (2002) studied the effects of known EDCs on the binding of *X. laevis* plasma thyroid hormone-binding proteins. Malathion was analysed as one of the known EDCs and showed weak inhibition of binding between triiodothyronine and *X. laevis* transthyretin, with no observable binding effects on thyroid hormone receptor-β. The results indicated target sites other than thyroid hormone receptors as playing a larger role in the thyroid-disrupting action of malathion. Ji et al. (2016) investigated the effects of malathion on *Xenopus* oocyte development and found that 100 μg/L significantly reduced the time to maturation. This effect was greatly increased by both 50 and 100 μg/L 1:1 mixtures between malathion and atrazine. The mixture of the two pesticides also increased the mortality rate significantly in oocytes that were fertilised post-exposure. However, de-jellying of eggs was observed to influence postfertilisation death rate which may have pronounced such effects.

3.5 Bioaccumulation Studies

3.5.1 Laboratory Studies

Pyrethroids

For the purposes of in vitro testing with transgenic *Xenopus* oocytes, Harril et al. (2005) investigated the link between dose and accumulation in non-transfected oocytes (normal *Xenopus* oocytes) to determine the active concentration of deltamethrin in in vitro studies. Their study showed deltamethrin as having the ability to accumulate actively in cell tissue in both a time-dependent and dose-dependent manner (Table 7). In exposed oocytes deltamethrin reached an accumulated concentration equivalent to exposure dose at 55.5 min. At 180 min the accumulated dose reached approximately double the media dose. The results from Harril et al. (2005) were further investigated by Watkins et al. (2007) who also showed a dose-dependent accumulation of deltamethrin in *X. laevis* oocytes after 1 h of exposure. Interestingly, voltage-sensitive sodium channel expressing (Na_v 1.2 + β_1)-transfected oocytes showed lower accumulation than non-transfected or sham-transfected oocytes. These sham-transfected oocytes underwent the transfection process as blanks being transfected only with water and thus having no genetic alteration. This was done in order to determine if the transfection method affected the oocyte response.

Oocytes, however, do not have the same enzymatic defences against xenobiotics as fully developed frogs do; thus this type of accumulation may not be seen in developed embryos. Although the ability of deltamethrin to accumulate in *Xenopus* oocytes proposes a possible risk to freshly laid frog eggs in the wild experiencing exposure events, it is also unknown how the stability of deltamethrin could be altered

Table 7 Laboratory-based bioaccumulation studies involving *Xenopus laevis* with regard to any of the 12 WHO-recommended MVCPs

Pesticide	Life stage	Exposure concentration (mg/L)	Exposure time (min)	Bioaccumulation concentration (μg/g)	Notes	Source
Deltamethrin	Unfertilised oocytes	0.5	180	1.08	Tissue levels increased with time reaching similar level to media dose at 55.5 min. Concentration-dependent increase in accumulation was also shown	Harril et al. (2005)
Deltamethrin	Unfertilised oocytes	0.05;0.5	60	0.98;0.62		Watkins et al. (2007)

when accumulated in tissue, perhaps extending the chemical's half-life in the process. Results from a field survey we performed (data not shown) to assess accumulation in adult *Xenopus muelleri* ($n = 12$) from a MVC sprayed area in South Africa collected in May 2016 (outside the MVC spraying season) showed no detectable accumulation of any MVCP (LOQ = 0.01 ppm) other than DDT in adult frogs, indicating that deltamethrin or other pyrethroid accumulation in wild *Xenopus*, if at all possible, is most likely temporary, with a quick cessation period, which reduces the risk of chronic exposure effects emerging.

3.5.2 Field Studies

Organochlorines

Only DDT is expected to show bioaccumulation in a field setting, as it is the only MVCP considered to be a persistent organic pollutant (POP). Three studies could be found on DDT accumulation in *Xenopus* sp. in the field (Table 8). Hothem et al. (2006) investigated the effects of pollutants on water birds from the Mojave Desert in North America (no active DDT use mentioned), incorporating the accumulation of pollutants in *X. laevis* into the study as a potential food source for birds. None of the frogs analysed had accumulated DDT or metabolite concentrations above limit of detection (0.01 μg/g), except one individual which had 0.054 μg/g wm *p,p*-DDE. Viljoen et al. (2016) measured DDT in *X. laevis* and *X. muelleri* fat bodies in an area sprayed with DDT for MVC through means of IRS in the Limpopo Province of South Africa. All DDT metabolites detected showed higher accumulation at sprayed sites compared to reference sites. No parent DDT (neither *p,p*-DDT nor *o,p*-DDT) was however detected in the frog fat bodies. The authors measured various morphometric factors which indicated observed testicular asymmetry but with no significant statistical differences. Total DDT levels measured in *Xenopus* sp. in this study ranged from 41 to 391 ng/g wet mass. The authors compared these values to literature on other frog species, and values for this study are at the lower end but within comparable range to concentrations measured in fat bodies of *Rhinella marina* in Mexico (Gonzalez-Mille et al. 2013) and of *Rana clamitans* in Canada (Harris et al. 1998). Wolmarans et al. (2018) measured DDT accumulation in whole body *X. muelleri* from another MVC-sprayed region in South Africa across four surveys over a 2-year period. The survey with the highest total DDT accumulation in this study showed a mean of 2,062 ng/g lipid. The study showed no discernible correlations between spraying periods and DDT accumulation. However, unlike the results of Viljoen et al. (2016), parent DDT was detected in two of the four surveys, with one survey showing a ratio of DDT:∑DDD/E (sum of metabolites) greater than one, indicative of exposure to new DDT. In Wolmarans et al. (2018) biomarker responses were also measured in liver and muscle tissue. The authors showed significant relationship between some of the measured biomarkers and both *p,p*-DDT and *p,p*-DDE, indicating that the accumulation of these two forms of DDT could cause changes in the energy dynamics of *X. muelleri*. The *p,p*-DDE

Table 8 Field-based bioaccumulation studies regarding *Xenopus* spp. and DDT

Pesticide	Species	Life stage	Tissue analysed	Measured concentration (ng/g wet mass)	Location (country)	Notes	Source
p,p-DDE	*Xenopus laevis*		Whole body	54.0	Edwards Air Force Base, Mojave Desert (USA)	Only one individual with detectable values	Hothem et al. (2006)
DDT (as ∑DDTs)	*Xenopus laevis*	Adult	Fat bodies	NS: 19.1–277.1 S: 5.5–914.4	Luvuvhu River, Limpopo province (South Africa)	NS = non-sprayed area; S = sprayed area. No *p,p*-DDT detected	Viljoen et al. (2016)
DDT (as ∑DDTs)	*Xenopus muelleri*	Adult	Fat bodies	NS: <LOQ–167.1 S: 154.8–658.1	Luvuvhu River, Limpopo province (South Africa)	NS = non-sprayed area; S = sprayed area. No *p,p*-DDT detected	Viljoen et al. (2016)
DDT (as ∑DDTs)	*Xenopus muelleri*	Adult	Whole body	<LOQ–79.59[a]	Pongola River floodplain, Northern KwaZulu-Natal (South Africa)	Liver and leg muscle sample removed before analysis	Wolmarans et al. (2018)

[a]Wolmarans et al. (2018) only reports lipid mass concentrations, but wet mass calculated values were obtained from the authors. Range shown includes all surveys

accumulation showed a reduction in acetylcholinesterase and malondialdehyde in the liver and energy availability in the form of carbohydrates and proteins in muscle tissue. The *p,p*-DDT accumulation indicated an increase in lipid energy availability and cellular energy consumption in muscle tissue.

Lambert et al. (1997, 2001) investigated frog deaths related to pesticide use and bioaccumulation in sub-Saharan Africa extensively but made no mention of *Xenopus* with regard to accumulation or deaths due to DDT or any of the other MVCPs. In this sense, the ecological resilience of *Xenopus* sp. is apparent, as population declines are rarely seen in areas where agricultural pesticides are regularly used. In this regard we hypothesise that some form of generational adaptation may result in the development of resistance to pesticides by wild populations. This is a factor that further brings into doubt the relevance of laboratory-based toxicity testing without some form of ecological context (Henry and Kishimba 2003).

3.6 *Environmental Concentrations*

To put into perspective the data discussed on *Xenopus*' sensitivity towards and accumulation of MVCPs, comparisons need to be made to environmental levels globally, with a focus on Africa as being the natural distribution of *Xenopus*.

3.6.1 Organochlorines

The highest environmental level of DDT measured in water in Africa was by Sibali et al. (2008) with total DDTs of 3.0 mg/L in Hartbeespoort Dam, South Africa (*p,p*-DDT was 1.5 mg/L). The highest recording previous to that was by Henry and Kishimba (2003) in Lake Victoria (Kenya) at 1.6 mg/L. Other African studies reported levels well below these levels with most values below 1 μg/L. Van Dyk et al. (2010) measured DDT in potable water from the Limpopo Province in South Africa (in an MVC area) at 7.6 μg/L of with the maximum *p,p*-DDT at 1 μg/L and maximum *o,p*-DDT at 0.8 μg/L. This is just below the *o,p*-DDT levels at which significant vitellogenin induction was shown (Palmer and Palmer 1995). However all of these levels are well above the sub-lethal threshold where changes in calling behaviour of *Xenopus laevis* were found (Hoffmann and Kloas 2016). The high water levels reported by Henry and Kishimba (2003) and Sibali et al. (2008) are also still well below acute toxicity levels and lower than the LOEC reported through FETAX studies showing no threat to *Xenopus* in their natural environment in terms of mortality. Most other African studies on DDT focused on sediment. The most recent study in Africa reporting water DDT concentration was by Ogbeide et al. (2018) who reported levels <0.001 ng/L total DDTs in surface waters from Edo State, Nigeria.

3.6.2 Pyrethroids

A recent review by Tang et al. (2018) summarises global environmental levels of pyrethroids who attributed their environmental distribution to agricultural use. Of the pyrethroids mentioned in the water dataset of Tang et al. (2018), cypermethrin is most frequently detected in water samples from around the world with the highest concentration reported in Bangladesh at 80.50 μg/L in lake water. Studies from China, the USA, and Czech Republic all showed levels close to or above 30 μg/L. These are levels that would lead to acute toxicity in *Xenopus* embryos and larvae according to the data previously discussed under acute toxicity. No African data are available on current pyrethroid levels in water. This makes comparison between *Xenopus*' sensitivity to pyrethroids less relevant as the environmental data are for regions outside of *Xenopus*' natural distribution. Historical levels of pesticides recorded in South African aquatic ecosystems were summarised by Ansara-Ross et al. (2012). Pyrethroids historically measured in South African waters include cyfluthrin (<0.0006 μg/L) and cypermethrin (up to 40.7 μg/L) in Northern KwaZulu-Natal province (an MVC region) (Sereda and Meinhardt 2003) and deltamethrin (1.43 μg/L) in runoff water (Dabrowski et al. 2002) in the Western Cape (not a MVC region). River water from the same Western Cape area (Louwrens River) where runoff was measured by Dabrowksi et al. (2002) was analysed by Bollmohr et al. (2007) who reported cypermethrin at 0.1 μg/L. Even though these are historical values from more than a decade ago, the maximum concentration of cypermethrin found in a MVC area by Sereda and Meinhardt (2003) was high enough to have theoretically caused acute toxicity to *Xenopus*. This region is host to both *X. laevis* and *X. muelleri*, but the latter is far more commonly found in this region today. Sensitivity differences between these species have never been assessed but may hypothetically be a contributing factor. In terms of sediment concentrations of pyrethroids, there are two studies from African countries outside of South Africa reporting pyrethroid concentrations in sediments. Daka et al. (2006) sampled sediment from different habitats in an area of the Okavango Delta in Botswana aerially sprayed with 0.26 g/ha deltamethrin for tsetse fly control. Deltamethrin in pool sediment had the highest levels which reached a maximum 5 days after spraying at 0.291 μg/g dry mass. This value dropped slightly towards 0.221 μg/g at 17 days after spraying. Olutona et al. (2016) found cyhalothrin at 77.75 μg/g dry mass in sediments from the Aiba stream in Iwo, Nigeria. This is extremely high considering Bollmohr et al. (2007) reported a maximum of 2.78 ng/g dry mass in sediments from the Western Cape in South Africa. Other historical sediment data from South Africa are reported in terms of wet mass concentrations making comparison difficult. Sereda and Meinhardt (2003) measured sediment deltamethrin levels up to 90 ng/g wet mass and cyfluthrin levels up to 467 ng/g wet mass in the MVC region of Northern KwaZulu-Natal.

3.6.3 Organophosphates

The only historical record of organophosphates in South African waters is malathion found at 20 ng/L in runoff at the Lourens River in the Western Cape Province (Thiere and Schulz 2004). Although the organic content from this agricultural runoff is higher and would be more likely to transport pesticides, pesticide concentrations are in the same range as levels found in headwaters of the Dongjiang River in China (malathion: 14.94–33.11 ng/L) by Chen et al. (2018). This is lower than the 0.7 nM (approximately 231.25 ng/L) IC50 for lysyl oxidase by malathion shown by Snawder and Chambers (1993).

Based on recorded levels of pesticides in African waters, pyrethroids (specifically cypermethrin) seem to pose the greatest risk to *Xenopus* in their natural environment. The data are, however, mostly outdated, and unfortunately there is no consistent ongoing monitoring of pesticide levels in African waters. The fact that both environmental levels and the combination of pesticide application use and dose data point to pyrethroids as a possible threat to *Xenopus* well-being in Africa warrants further investigation into actual environmental pyrethroid and pesticide mixtures levels of exposure to all frogs and not *Xenopus*.

4 Future Prognosis for Amphibian Ecotoxicology

Based on the data reviewed, it is clear that gaps exist within amphibian, specifically anuran, ecotoxicology literature. Acute toxicity data for one of the standard test species was available for only half of the WHO-recommended pesticides, and not enough data on different species exists for species sensitivity distribution analyses. In addition, most of the available acute toxicity data can be considered old literature. Ethics in vertebrate toxicological research may well be a contributing reason behind this lack of data. However, considering comments by Lillicrap et al. (2016) on the future of vertebrate ecotoxicology and steps forward, standardised tests such as FETAX, AMA, and LAGDA – mentioned specifically as the “go to” current methods for amphibian ecotoxicology – should have minimal issues associated with them in terms of ethical approval as these are globally accepted standardised tests that are also used as alternatives to other animal tests. A recent publication by Brod et al. (2018) does suggest that amphibian research is at a crossroads in terms of animal welfare in research. If questions posed by Brod et al. (2018) on amphibian welfare (such as appropriate euthanasia methods and measurable biomarkers of welfare applicable to amphibians) aren’t sufficiently answered in the near future, along with renewed interest in amphibian research by funders, it may greatly hinder progress made in generating relevant and usable ecotoxicology data on amphibians.

Another possible reason for the lack of data already existing in this field is a lack of interest amongst researchers. Even though the amphibian ecotoxicology research community around the world is quite small, amphibians are one of the fastest-vanishing groups in the world, which should be an incentive for more detailed research. Zippel and Mendelson (2008) wrote a call to action on the global amphibian crisis (i.e. the rapid loss of amphibian species globally). The authors state that a valuable lesson from this crisis is the way in which the, then, 20 years of monitoring provided increasing knowledge on amphibian declines but did little to prevent extinctions globally and the term "business as usual" is deemed no longer sufficient by itself for resolving environmental issues. In our opinion, a decade later, many of the questions posed have still not been answered adequately. There is a clear need not only for toxicity or sensitivity data of anurans towards MVCPs but also linkages between application rates and exposure and how these relate to accumulation, or effects, as well as linkages between laboratory and real-world scenarios and monitoring of current environmental exposure levels. The exploration of mixture effects is something that has only started to progress recently in amphibian ecotoxicology and could prove useful in amphibian health risk assessment studies in future. There is a need for comprehensive datasets on amphibian ecotoxicology, which can be achieved only through inter-laboratory collaboration and data-sharing in this regard.

5 Conclusion

The clearest issue with anuran ecotoxicology regarding pesticides is consistency between reports, mostly due to method-related differences. This is partly accounted for by the fairly common use of FETAX. But the second limiting factor, availability of data, then comes into play. Very few records exist regarding the standard test species, *X. laevis*, and casting an even wider net by including the whole genus *Xenopus* does not increase the availability by any significant means. It is of concern, since most of these pesticides are still being used in agriculture and sectors other than MVC around the world, that for the 12 WHO-recommended MVCPs so little data is available. Biomonitoring data of these pesticides in Africa is also outdated.

Based on the available literature, of the four pesticide groups, pyrethroids may hold the greatest acute toxicity potential to *Xenopus*. This does not mean that at current usage, *Xenopus* are at risk of acute toxicity, but merely that, if any of the MVCPs were to result in acute toxicity, based on the sensitivity of *Xenopus* towards pyrethroids and current and recommended usage, pyrethroids would be the most likely compound group to show such effects in the field; however, these potentials are based solely on the data available and are not definitive values in a broader context than how they are used in this study. Based on sub-lethal activity, it is clear that very little is known about how these pesticides affect anurans, in particular at sub-lethal concentrations. The field data available speaks to the persistence of DDT in the environment. As DDT is the only one of the MVCPs known to bioaccumulate in frogs, of the 12 MVCPs, it can be considered to carry the greatest chronic risk to all anurans for this reason. The EDC effects of *p,p*-DDE on *Xenopus* calling behaviour raise concern, as *p,p*-DDE is the most persistent metabolite of DDT and shows the greatest bioaccumulation potential. If more sub-lethal effects linking with accumulation, such as those measured by Wolmarans et al. (2018), can be confirmed through laboratory and field testing, it will increase the certainty of risk values that can be calculated from field accumulation data for amphibian populations.

6 Recommendations

There is a definite need for further monitoring of anurans in Africa with regard to the effects of pesticides, specifically within the genus *Xenopus*. Lambert (2001) stated the need for using African amphibians, such as *Xenopus* in particular, more extensively as a monitoring tool in Africa. Increased use of *Xenopus* in monitoring is not reflected by literature as of yet. Based on the aim of this review, the authors encourage new research on anuran ecotoxicology, specifically in terms of insecticides used in vector control and agriculture, in order to fill the gaps in the current knowledge base. In future studies, *Xenopus* sp. could be used as a starting point around which to build a comprehensive toxicological dataset in order better to understand the impact that human activity has on anuran well-being. The use of adverse outcome pathways may serve as a useful tool to understand sub-lethal effects observed and is yet to be used in anuran studies. However, the basic empirical toxicity of these pesticides towards anurans still requires attention.

7 Summary

This review concerns the genus *Xenopus* and literature relating it to malaria vector control pesticides. Criteria for literature to be included for review include that dealing with *Xenopus* sp. and exposure to any one of the 12 WHO-recommended MVCPs. The data analysed showed that, with the available acute toxicity literature, the pyrethroid pesticide group holds the greatest potential to cause acute toxicity when used for malaria vector control, but acute toxicity cannot be interpreted as a likely outcome based on the data available and limitations to the analyses performed. This potency is increased by the concurrent use of these pesticides in agriculture at even greater doses. The only persistent pollutant in the list of pesticides analysed is DDT. Field literature show that DDT accumulates (actively, in some cases) in *Xenopus* sp. in a natural setting both inside and outside MVC spraying areas, with possible sub-lethal effects associated with the accumulated concentrations. There is a lack of published research in terms of both acute toxicity and sub-lethal effects with regard to anurans in general and MVCPs. Some of the recommended pesticides have no associated toxicity literature with regard to any anuran standard test species, let alone to *Xenopus*. This review serves to indicate the need for more detailed information regarding pesticides recommended for vector control and for anurans in general, both in a laboratory and field setting. With amphibians declining rapidly globally and with *Xenopus* being a well-established standard test species, this species may serve as a good starting point for building a comprehensive dataset in this regard.

Acknowledgements This study was partially funded by the Flemish Interuniversity Council (VLIR) to ECN (VLIR-OUS project – ZEIN21013PR396), and financial assistance was also provided by the South African National Research Foundation (NRF: Grant no. SFH150624120779). Opinions expressed and conclusions arrived at are those of the authors and are not necessarily to be attributed to VLIR or the NRF.

The authors declare that they have no conflict of interest.

References

Akkermans LMA, Van den Bercken J, Versluijs-Helder M (1975) Comparative effects of DDT, allethrin, and aldrin-transdiol on sense organs of *Xenopus laevis*. Pest Biochem Physiol 5:451–457

Alonso P, Noor AM (2017) The global fight against malaria is at crossroads. Lancet 390:2532–2534

Ansara-Ross TM, Wepener V, Van den Brink PJ, Ross MJ (2012) Pesticides in South African fresh waters. Afr J Aquat Sci 37:1–16

Århem P, Frankenhaeuser B (1974) DDT and related substances: effects on permeability properties of myelinated *Xenopus* nerve fibre. Potential clamp analysis. Acta Physiol Scand 91:502–511

Århem P, Frankenhaeuser B, Göthe R, O'Bryan P (1974) DDT and related substances on myelinated nerve: effects on permeability properties. Acta Physiol Scand 91:130–132

ASTM (American Society for Testing and Materials) (2012) Standard guide for conducting the frog embryo teratogenesis assay – *Xenopus* (FETAX). Annual book of ASTM standards. American Society for Testing and Materials, Philadelphia

ATSDR (Agency for Toxic, Substances and Disease Registry) (2002) Toxicological profile for DDT, DDE, and DDD. Agency for Toxic, Substances and Disease Registry. U.S. Department of Health and Human Services, Atlanta. http://www.atsdr.cdc.gov/toxprofiles/tp35.pdf. Accessed Aug 2018

Avila VL, Frye PG (1978) Feeding behaviour of the African clawed frog (*Xenopus Laevis* Daudin): (Amphibia, Anura, Pipidae): effect of prey type. J Herpetol 12:391–396

Aydin-Sinan H, Güngördü A, Ozmen M (2012) Toxic effects of deltamethrin and λ-cyhalothrin on *Xenopus laevis* tadpoles. J Environ Sci Health B 47:397–402

Bollmohr S, Day JA, Schulz R (2007) Temporal variability in particle associated pesticide exposure in a temporarily open estuary, Western Cape, South Africa. Chemosphere 68:479–488

Bonfanti P, Colombo A, Orsi F, Nizzetto I, Andrioletti M, Bachetta R, Mantecca P, Fascio U, Vailati G, Vismara C (2004) Comparative teratogenicity of chlorpyifos and malathion on *Xenopus laevis* development. Aquat Toxicol 70:189–200

Bouwman H, Van den Berg H, Kylin H (2011) DDT and malaria prevention: addressing the paradox. Environ Health Perspect 119:744–747

Brod S, Brooks L, Garner TWJ (2018) Discussing the future of amphibians in research. Lab Anim. https://doi.org/10.1038/s41684-018-0193-6

Burggren WW, Warburton S (2007) Amphibians as animal models for laboratory research in physiology. ILAR J 48:260–269

Channing A (1998) Tadpoles as bio-indicators of stream quality: a baseline study. Water Research Commission of South Africa (WRC) Technical Report No. 718/1/98

Chemotti DC, Davis SN, Cook LW, Willoughby IR, Paradise CJ, Lom B (2006) The pesticide malathion disrupts *Xenopus* and Zebrafish embryogenesis: an investigative laboratory exercise in developmental toxicology. Biocene 32:3–18

Chen Y, Yua K, Hassan M, Xu C, Zhang B, Gin KY, He Y (2018) Occurrence, distribution and risk assessment of pesticides in a river reservoir system. Ecotox Environ Safe 166:320–327

Dabrowski JM, Peall SKC, Reinecke AJ, Liess M, Schulz R (2002) Runoff-related pesticide input into the Lourens River, South Africa: basic data for exposure assessment and risk mitigation at the catchment scale. Water Air Soil Pollut 135:265–283

Daka PS, Obuseng VC, Torto N, Huntsman-Mapila P (2006) Deltamethrin in sediment samples of the Okavango Delta, Botswana. Water SA 32:483–488

Dalu T, Weyl OLF, Froneman PW, Wasserman RJ (2016) Trophic interactions in an austral temperate ephemeral pond inferred using stable isotope analysis. Hydrobiologia 768:81–94

Dumont JN, Schultz TW, Buchanan MV, Kao GL (1983) Frog embryo teratogenesis assay: *Xenopus* (FETAX) – a short-term assay applicable to complex environmental mixtures. In: Waters MD, Sandhu SS, Lewtas J, Claxton L, Chernoff N, Nesnow S (eds) Short-term bioassays in the analysis of complex environmental mixtures III. Environmental science research, vol 27. Springer, Boston, pp 393–405

Elliot-Feeley E, Armstrong JB (1982) Effects of fenitrothion and carbaryl on *Xenopus laevis* development. Toxicology 22:319–335

FEOW (Freshwater Ecoregions of the World) (2008) Freshwater amphibian species richness map. http://www.feow.org/maps/biodiversity/freshwater_amphibian_species_richness. Accessed Sept 2018

Glunt KD, Blanford JI, Paaijmans KP (2013) Chemicals, climate, and control: increasing the effectiveness of malaria vector control tools by considering relevant temperatures. PLoS Pathog 9:e1003602

Gonzalez-Mille DJ, Espinosa-Eyes G, Rivero-Péres NE, Trejo-Acevedo A, Nava-montes AD, Ilizaliturri-Hernández CA (2013) Persistent organochlorine pollutants (POPs) and DNA damage in Giant Toads (*Rhinella marina*) from an industrial area at Coatzacoalcos. Mexico Water Air Soil Pollut 224:1781

Gosner KL (1960) A simplified table for staging anuran embryos and larvae with notes on identification. Herpetologica 16:183–190

Gurdon JB, Hopwood N (2000) The introduction of *Xenopus laevis* into developmental biology: of empire, pregnancy testing and ribosomal genes. Int J Dev Biol 44:43–50

Harril JA, Meacham CA, Shafer TJ, Hughes MF, Crofton KM (2005) Time- and concentration-dependent accumulation of [^{3}H]-deltamethrin in *Xenopus laevis* oocytes. Toxicol Lett 157:79–88

Harris M, Bishop CA, Struger J, Van Den Heuvel MR, Van Der Kraak GJ, Dixon DG, Ripley B, Bogart JP (1998) The functional integrity of northern leopard frog (*Rana pipiens*) and green frog (*Rana clamitans*) populations in orchard wetlands I. Genetics physiology, and biochemistry of breeding adults and young-of-the-year. Environ Toxicol Chem 17:1338–1350

Henry L, Kishimba MA (2003) Levels of pesticide residues in water, soil and sediments from southern Lake Victoria and its basin. Tanz J Sci 29:77–89

Hoffmann F, Kloas W (2016) *p,p′*-Dichlorodiphenyldichloroethylene (*p,p′*-DDE) can elicit antiandrogenic and estrogenic modes of action in the amphibian *Xenopus laevis*. Physiol Behav 167:172–178

Hothem RL, Crayon JJ, Law MA (2006) Effects of contaminants on reproductive success of aquatic birds nesting at Edwards Air Force Base, California. Ach Environ Contam Toxicol 51:711–719

Hu L, Zhu J, Rotchell JM, Wu L, Gao J, Shi H (2015) Use of the enhanced frog embryo teratogenesis assay-*Xenopus* (FETAX) to determine chemically-induced phenotypic effects. Sci Total Environ 508:258–265

Ji Q, Lee J, Lin Y, Jing G, Tsai LJ, Chen A, Hetrick L, Jocoy D, Liu J (2016) Atrazine and malathion shorten the maturation process of *Xenopus laevis* oocytes and have an adverse effect on early embryo development. Toxicol In Vitro 32:63–69

Knaak JB, Dary CC, Zhang X, Gerlach RW, Tornero-Velez R, Chang DT, Goldsmith R, Blancato JN (2012) Parameters for pyrethroid insecticide QSAR and PBPK/PD models for human risk assessment. Rev Environ Contam Toxicol 219:1–114

Lambert MRK (1997) Effects of pesticides on amphibians and reptiles in sub-Saharan Africa. Rev Environ Contam Toxicol 150:31–73

Lambert MRK (2001) Death from pesticides reviewed among non-target amphibians in sub-Saharan Africa. Herpetol Bull 78:21–27

Lillicrap A, Belanger S, Burden N, Du Pasquier D, Embry MR, Halder M, Lampi MA, Lee L, Norberg-King T, Rattner BA (2016) Alternative approaches to vertebrate ecotoxicity tests in the 21st century: a review of developments over the last two decades and current status. Environ Toxicol Chem 35:2637–2646

Lindholm M, Hesen DO, Mosepele K, Wolski P (2007) Food web and energy fluxes on a seasonal floodplain: the influence of flood size. Wetlands 27:775–784

Lutz I, Kloas W (1999) Amphibians as a model to study endocrine disruptors: I. Environmental pollution and estrogen receptor binding. Sci Total Environ 225:49–57

Mann RM, Hyne RV, Choung CB, Wilson SP (2009) Amphibians and agricultural chemicals: review of the risks in a complex environment. Environ Pollut 157:2903–2927

MAP (Malaria Atlas Project) (2018) Temperature suitability index for Plasmodium vivax/ falciparum transmission. https://map.ox.ac.uk/explorer/#/explorer. Accessed Sept 2018

Martini F, Fernández C, Segundo LS, Tarzona JV, Pablos VM (2010) Assessment of potential immunotoxic effects caused by cypermethrin, fluoxetine, and thiabendazole using heat shock protein 70 and interleukin-1β mRNA expression in the anuran *Xenopus laevis*. Environ Toxicol Chem 29:2536–2543

Nieuwkoop PD, Faber J (1994) Normal table of *Xenopus laevis* (daudin): a systematical & chronological survey of the development from the fertilized egg till the end of metamorphosis. Garland Science, Oxford

Ogbeide O, Chukwuka A, Tongo I, Ezemonye L (2018) Relationship between geosorbent properties and field-based partition coefficients for pesticides in surface water and sediments of selected agrarian catchments: implications for risk assessment. J Environ Manag 217:23–17

Olutona GO, Olatunji SO, Obisanya JF (2016) Downstream assessment of chlorinated organic compounds in the bed-sediment of Aiba Stream, Iwo, South-Western, Nigeria. Springerplus 5:67. https://doi.org/10.1186/s40064-016-1664-0

Palmer BD, Palmer SK (1995) Vitellogenin induction by xenobiotic estrogens in the red-eared turtle and African clawed frog. Environ Health Perspect 103:19–25

Pan D, Liang X (1993) Safety study of pesticide on Bog Frog, a predatory natural enemy of pest in paddy field. J Hunan Agric College 19:47–54

Pauli BD, Perrault JA, Money SL (2000) RATL: a database of reptile and amphibian toxicology literature. Technical Report Series No. 357. Canadian Wildlife Service. Québec, Canada

Robert J, Ohta Y (2009) Comparative and developmental study of the immune system in *Xenopus*. Dev Dyn 238:1249–1270

Rudek Z, Rożek M (1992) Induction of micronuclei in tadpoles of *Rana temporaria* and *Xenopus laevis* by the pyrethroid Fastac 10 EC. Mutat Res 298:25–29

Ruigt GSF, Van den Bercken J (1986) Action of pyrethroids on a nerve-muscle preparation of the Clawed Frog, *Xenopus laevis*. Pest Biochem Physiol 25:176–187

Saka M (2004) Developmental toxicity of *p,p'*-dichlorodiphenyltrichloroethane, 2,4,6-trinitrotoluene, their metabolites, and benzo[*a*]pyrene in *Xenopus laevis* embryos. Environ Toxicol Chem 23:1065–1073

Sereda BL, Meinhardt HR (2003) Insecticide contamination of the water environment in malaria endemic areas of KwaZulu-Natal (SA): the risk of insecticide resistance development in malaria vectors. WRC Report No. 1119/1/03. Pretoria: Water Research Commission

Sibali LL, Okwonkwo JO, McCrindle RI (2008) Determination of selected organochlorine pesticide (OCP) compounds from the Jukskei River catchment area in Gauteng. South Africa Water SA 34:611–622

Snawder JE, Chambers JE (1989) Toxic and developmental effects of organophosphorus insecticides in embryos of the South African Clawed Frog. J Environ Sci Health B 24:205–218

Snawder JE, Chambers JE (1990) Critical time periods and the effect of tryptophan in malathion-induced developmental defects in *Xenopus laevis*. Life Sci 46:1635–1642

Snawder JE, Chambers JE (1993) Osteolathyrogenic effects of malathion in *Xenopus* embryos. Toxicol Appl Pharmacol 121:210–216

Tang W, Wang D, Wang J, Wu Z, Li L, Huang M, Xu S, Yan D (2018) Pyrethroid pesticide residues in the global environment: an overview. Chemosphere 191:990–1007

Thiere G, Schulz R (2004) Runoff-related agricultural impact in relation to macroinvertebrate communities of the Lourens River, South Africa. Water Res 38:3092–3102

UNEP (2009) Stockholm Convention on Persistent Organic Pollutants (POPs), text and annexes as amended in 2009. https://www.wipo.int/edocs/lexdocs/treaties/en/unep-pop/trt_unep_pop_2.pdf. Accessed Feb 2019

Van den Bercken J, Akkermans LMA, Van der Zalm JM (1973a) DDT-like action of allethrin in the sensory nervous system of *Xenopus laevis*. Eur J Pharmacol 21:95–106

Van den Bercken J, Akkermans LMA, Van Langen RG (1973b) The effect of DDT and dieldrin on skeletal muscle fibres. Eur J Pharmacol 21:89–94

Van der Oost R, Beyer J, Vermeulen NPE (2003) Fish bioaccumulation and biomarkers in environmental risk assessment: a review. Environ Toxicol Chem 13:57–149

Van Dyk JC, Bouwman H, Barnhoorn IEJ, Bornman MS (2010) DDT contamination from indoor residual spraying for malaria control. Sci Total Environ 408:2745–2752

Vijverberg HPM, Van den Bercken J (1979) Frequency-dependent effects of the pyrethroid insecticide decamethrin in frog myelinated nerve fibres. Eur J Pharmacol 58:501–504

Vijverberg HPM, Van den Bercken J (1982) Action of pyrethroid insecticides on the vertebrate nervous system. Neuropathol Appl Neurobiol 8:421–440

Vijverberg HPM, Ruigt GSF, Van den Bercken J (1982a) Structure-related effects of pyrethroid insecticides on the lateral-line sense organ and on peripheral nerves of the Clawed Frog, *Xenopus laevis*. Pest Biochem Physiol 18:315–324

Vijverberg HPM, Van der Zalm JM, Van den Bercken J (1982b) Similar mode of action of pyrethroids and DDT on sodium channel gating in myelinated nerves. Nature 295:601–603

Vijverberg HPM, Van der Zalm JM, Van Kleef RGDM, Van den Bercken J (1983) Temperature- and structure-dependent interaction of pyrethroids with the sodium channels in frog node of Ranvier. Biochim Biophys Acta 728:73–82

Viljoen IM, Bornman R, Bouwman H (2016) DDT exposure to frogs: a case study from Limpopo Province, South Africa. Chemosphere 159:335–341

Walshe DP, Garner P, Adeel AA, Pyke GH, Burkot TR (2017) Larvivorous fish for preventing malaria transmission. Cochrane Database Syst Rev 12:CD008090

Watkins JA, Mecham CA, Crofton KM, Shafer TJ (2007) Concentration-dependent accumulation of [^{3}H]-deltamethrin in sodium channel $Na_v1.2/\beta_1$ expressing *Xenopus laevis* oocytes. Toxicol In Vitro 21:1672–1677

Webb C, Crain DA (2006) Effects of ecologically relevant doses of malathion on developing *Xenopus laevis* tadpoles. Bios 77:1–6

WHO (World Health Organization) (1971) Alternative insecticides for vector control. World Health Org Bull 44:1–446

WHO (World Health Organization) (2006) Indoor residual spraying: use of indoor spraying for scaling up global malaria control and elimination. http://whqlibdoc.who.int/hq/2006/WHO_HTM_MAL_2006.1112_eng.pdf. Accessed Aug 2018

Wolmarans NJ, Wepener V, Du Preez LH, Ikenaka Y, Ishizuka M, Smit NJ (2015) Assessment of the food web structure of *Xenopus muelleri* from the Phongolo River floodplain using stable isotope analysis. Peer reviewed conference proceedings of the 7th international toxicology symposium in Africa; 31 August 2015, Johannesburg, South Africa

Wolmarans NJ, Du Preez LH, Yohannes YB, Ikenaka Y, Ishizuka M, Smit NJ, Wepener V (2018) Linking organochlorine exposure to biomarker response patterns in anurans: a case study of Müller's clawed frog (*Xenopus muelleri*) from a tropical malaria vector control region. Ecotoxicology 27:1203–1216

Yamauchi K, Eguchi R, Shimada N, Ishihara A (2002) The effects of endocrine-disrupting chemicals on thyroid hormone binding to *Xenopus laevis* transthyretin and thyroid hormone receptor. Clin Chem Lab Med 40:1250–1256

Yu S, Wages M, Cai Q, Maul JD, Cobb GP (2013) Lethal and sublethal effects of three insecticides on two developmental stages of *Xenopus laevis* and comparison with other amphibians. Environ Toxicol Chem 32:2056–2064

Yu S, Weir SM, Cobb GP, Maul JD (2014) The effects of pesticide exposure on ultraviolet-B radiation avoidance behavior in tadpoles. Sci Total Environ 481:75–80

Yu S, Tang S, Mayer GD, Cobb GP, Maul JD (2015a) Interactive effects of ultraviolet-B radiation and pesticide exposure on DNA photo-adduct accumulation and expression of DNA damage and repair genes in *Xenopus laevis* embryos. Aquat Toxicol 159:256–266

Yu S, Wages M, Willming M, Cobb GP, Maul JD (2015b) Joint effects of pesticides and ultraviolet-B radiation on amphibian larvae. Environ Pollut 207:248–255

Zippel KC, Mendelson JR (2008) The amphibian extinction crisis: a call to action. Herpetol Rev 39:23–29

The manufacturer's authorised representative in the EU is Springer Nature Customer Service Centre GmbH, Europaplatz 3, 69115 Heidelberg, Germany. If you have any concerns regarding our products, please contact ProductSafety@springernature.com

Printed and bound by CPI Group (UK) Ltd, Croydon, CR0 4YY
05/10/2026
02228156-0007